T0192833

Emerging Topics in Statistics and Biostatistics

More information about this series at http://www.springer.com/series/16213

Ding-Geng (Din) Chen • Jenny K. Chen

Statistical Regression Modeling with *R*

Longitudinal and Multi-level Modeling

 Springer

Ding-Geng (Din) Chen
School of Social Work and
Gillings School of Global Public Health
University of North Carolina at Chapel Hill
Chapel Hill, NC, USA

Jenny K. Chen
Department of Statistics
Cornell University
Ithaca, NY, USA

ISSN 2524-7735   ISSN 2524-7743   (electronic)
Emerging Topics in Statistics and Biostatistics
ISBN 978-3-030-67585-1   ISBN 978-3-030-67583-7   (eBook)
https://doi.org/10.1007/978-3-030-67583-7

This Springer imprint is published by the registered company Springer Nature Switzerland AG
The registered company address is: Gewerbestrasse 11, 6330 Cham, Switzerland

Preface

This book is originated from many years of research and teaching of this course to master's and doctoral-level students as an applied statistical course in statistical regression, with a focus on multi-level modeling. It has now been created as a textbook and a reference book for future teaching and learning.

Why Read This Book

Most researchers and practitioners, especially in social sciences and public health, use data that have multi-level structure or are clustered to inform research and policy with the goal of improving population well-being. Therefore, it is paramount to produce valid and efficient estimates from these data. In most instances, if not all, classical regression modeling is not suited for the analysis of multi-level or clustered data to produce valid and efficient estimates because of the correlations that existed in the multi-level data. Instead, we need to use advanced multi-level modeling methods to address these limitations of classical regression analysis. This book is designed to serve as introductory and practical guide to analyzing multi-level data. Specifically, this book serves as the transition from the classical regression modeling, such as multiple linear regression, logistic regression, Poisson regression, generalized linear regression, and nonlinear regression, to the more advanced multi-level modeling due to the multi-level data structure from research and applications.

Multi-level data structure is typically referred to as nested data structure or clustered data structure. For example, if you want to examine the impact of a new teaching intervention on student performance within or across educational systems and geographical areas, it might be impractical logistically and financially to randomly sample students at the individual level in a randomized controlled intervention design. However, we could use a cluster-randomized design and randomize at the *classroom* level, which are not only practical but will also give you a valid estimate of teaching intervention on student performance. In this situation, the *classrooms* act as clusters based on the teachers in each classroom. Each cluster (i.e., classroom) will then be randomly assigned to either a treatment or a control condition. Hence, if a specific classroom is assigned to the treatment/control condition, all students in that classroom would receive the new teaching intervention. Due to clustering, students within the classrooms are dependent on each other due to the teaching from the same teacher. Therefore, the independence assumption from the classical regression models is violated, and the model estimation could be then biased if the classical regression modeling is used.

This type of data structure is different from the data structure from all the classical regression models where data are required to be independent. With such multi-level data structure, new techniques are needed to extend classical regression models. This book is then written to introduce multi-level data analysis, i.e., multi-level modeling or multi-level regression, extending from the classical regression methods and longitudinal data analysis.

There are different terminologies for *multi-level data*. Readers may find others using *nested data* or *clustered data* or *hierarchical data*. Similarly, *multi-level modeling* may be referred as *multi-level regression, hierarchical regression*, or *hierarchical modeling* in most social science literature. Multi-level modeling is called *mixed-effects* modeling in statistics since it is a modeling technique with a mixture of random-intercepts and random-slopes. Multi-level modeling will be used in this book and denoted as *MLM* along with the typical notation of *MLR* in multiple linear regression.

Structure of the Book

This book is structured in ten chapters with datasets and *R* packages introduced in Appendices. Chapters 1 to 7 are for data analysis with continuous outcome in multi-level data structure (i.e., the so-called normally distributed data with nested structure) and their implementation using the *R* package *nlme*. Specifically, Chap. 1 is to provide an overview of the classical linear regression, which covers simple linear regression and multiple linear regression. This is designed to lay the foundation to multi-level linear regression in the future chapters. Chapter 2 then extends Chap. 1 with an introduction to the concepts of multi-level linear regression. Chapter 3 discusses the detailed analysis of two-level multi-level regression, followed by Chap. 4 on three-level multi-level regression. Since longitudinal data can be easily structured as multi-level data, Chap. 5 then discusses longitudinal data analysis. Extending linear regression to nonlinear regression, Chap. 6 introduces the techniques of nonlinear regression that focus on logistic growth modeling, followed by Chap. 7 on nonlinear multi-level regression.

Chapters 8 to 10 are for data analysis with a non-normal outcome (we focus on *binary* and *counts* data) using *R* package *lme4*. Specifically, Chap. 8 is to review the logistic regression for categorical data, Poisson regression, and negative-binomial regression for counts data along with the generalized linear regression. Chapter 9 is to discuss generalized logistic regression with multi-level categorical data, followed by Chap. 10 on generalized Poisson regression for multi-level counts data.

All the datasets used in the book are detailed in Appendix A. The associated *R* packages *nlme* and *lme4* for MLM in this book are discussed in Appendix B.

Chapel Hill, NC, USA   Ding-Geng (Din) Chen

Ithaca, NY, USA   Jenny K. Chen

Acknowledgments

We owe a great deal of gratitude to many who helped in the completion of the book. We have been fortunate enough to work with a number of Ph.D. students (Anderson Al Wazni, Ishrat Alam, Theo Beltran, Alpha Oumar Diallo, Mekhala Dissanayake, Josée Dussault, Daniel Gibbs, Terence Johnson, Emma Rosen, Maria Stevens, Alberto Valido Delgado, and Erica Zeno) who proofread the book. Special thanks to Melissa R. Jenkins who helped greatly in preparing data analysis, and Professor Nancy A. Quick who carefully proofread the book. We also gratefully acknowledge the professional support of Ms. Laura Aileen Briskman and her team from Springer, who made the publication of this book a reality. To everyone involved in the making of this book, we say **Thank You!**

Thanks also go to the School of Social Work, University of North Carolina-Chapel Hill, for its support, and the continual encouragement from Dean Gary L. Bowen on this project. This work on the book was greatly facilitated by a research sabbatical to the first author (Ding-Geng Chen) from the School of Social Work.

This book is written using *bookdown* package (Xie, 2019), and we thank the package creator, Dr. Yihui Xie, to have created such a wonderful package for public use. We also thank the *R* creators and the community for creating a wonderful open-source *R* computing environment, so that this book can be possible. **Thank You All!**

Chapel Hill, NC, USA   Ding-Geng (Din) Chen

Ithaca, NY, USA   Jenny K. Chen

Contents

About the Authors

Ding-Geng (Din) Chen is a fellow of the American Statistical Association, an elected member of the International Statistical Institute, and a fellow of the Society of Social Work and Research. Currently, he is the Wallace H. Kuralt distinguished professor at the School of Social Work with a joint professor in Biostatistics at the Department of Biostatistics, Gillings School of Global Public Health, University of North Carolina at Chapel Hill, USA. He was a professor in Biostatistics at the University of Rochester and the Karl E. Peace endowed eminent scholar chair in biostatistics at the Georgia Southern University. He is also a senior statistics consultant for biopharmaceuticals and government agencies with extensive expertise in Monte Carlo simulations, clinical trial biostatistics, and public health statistics. Dr. Chen has more than 200 referred professional publications, co-authored and co-edited 30 books on clinical trial methodology and analysis, meta-analysis, data sciences, causal inferences, and public health applications. He has been invited nationally and internationally to give speeches on his research.

Jenny K. Chen graduated with a master's degree at the Department of Statistics and Data Science at Cornell University. She is currently working as a financial analyst at the Morgan Stanley (Midtown New York) for their Wealth Management division. Previously, Jenny has worked as a product manager for Google, where she led a team of data scientists to develop several prediction algorithms for the 2019 NCAA March Madness Basketball Tournament. She has also published several research papers in statistical modeling and data analytics.

Chapter 1
Linear Regression

We start this book with an overview of linear regression modeling. This is to bring all readers to the same level of understanding of regression modeling before we introduce multi-level modeling (MLM).

We illustrate the regression technique using the *WHO Statistics on Life Expectancy* data detailed in Appendix A.1 as an example, with the purpose of identifying factors that may be associated with life expectancy. A step-by-step illustration will be implemented with *R* to analyze this data with both simple linear regression (*SLR*) and multiple linear regression (*MLR*).

Note to readers: We will use *R* package *car* (i.e., Companion to Applied Regression) in this chapter. Remember to install this *R* package to your computer using *install.packages("car")* and load this package into *R* session using *library(car)* before running all the *R* programs in this chapter.

1.1 Descriptive Data Analysis

1.1.1 Data and Summary Statistics

We first read the data into *R*. Since this dataset is saved in the *.csv* file format, we can read the dataset into *R* using *read.csv* as follows:

```
# read the data into R session using 'read.csv'
dWHO = read.csv("WHOLifeExpectancy.csv", header=T)
# Check the data dimmension
dim(dWHO)
```

```
## [1] 2938    22
```

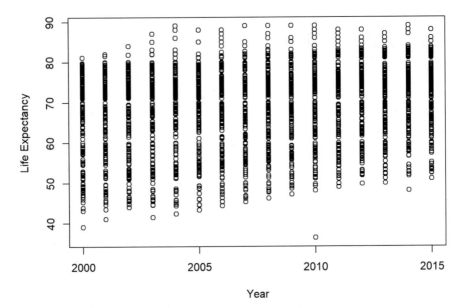

Fig. 1.1 Illustration of life expectancy over time

As seen from the output, this dataset contains 2938 observations and 22 variables. We can summarize the data as shown in Sect. A.1. From the summary output, we can see the data distribution for all variables in the data. As seen from the summary output, the continuous variables will be summarized as the *min.* (i.e., the minimum value), *1st Qu.* (i.e., the first quantile), *Median* (i.e., the data median), *Mean* (i.e., the mean value), *3rd Qu.* (i.e., the third quantile), and *Max.* (i.e., the maximum value). For categorical variables, the frequencies will be summarized. For all variables, the number of missing values (i.e., *NA's*) will be printed.

1.1.2 Preliminary Graphical Analysis

Typical preliminary analysis consists of plotting the data to identify whether or not a potential linear relationship exists between the outcome and independent variables.

Let us consider a simple scenario with the *WHO* data in which we examine the changes of life expectancy over time for all countries. We would plot the data as seen in Fig. 1.1 as follows:

```
plot(LifeExpectancy~Year,xlab="Year",ylab="Life Expectancy", dWHO)
```

As seen from this Fig. 1.1, the life expectancy generally increases from year 2000 to 2015 for all countries together, which is approximately linear even though there exists a large variation.

Keep in mind that data graphics are enormously helpful for understanding the regression models behind the mathematical equations. Intuitively we know that life expectancy has been increasing over time, and seeing this in Fig. 1.1 puts it into clearer perspective. From this figure, we can also see how much noise (i.e., large variations in this linear trend) this data has. This leads us to the perspective for understanding why we need to control for variations with extra variables for moderation analysis and multiple regression analyses.

1.2 Review of Multiple Linear Regression Models

1.2.1 Multiple Linear Regression Model

Before we analyze any real data, let us briefly review the regression models. Suppose that we have a dataset with N observations (for example, $N = 2938$ in the *dWHO* dataset) from participants in a study with outcome variable y, for example, the variable *LifeExpectancy* in Fig. 1.1, which is related and predicted from p independent variables of $x_1, \cdots, x_p$. The general multiple linear regression (*MLR*) model can be written in the following equation:

$$y_i = \beta_0 + \beta_1 x_{1i} + \cdots + \beta_p x_{pi} + \epsilon_i \qquad (1.1)$$

where $i = 1, \cdots, N$ for the ith observation.

For simplicity, we can put the MLR model in Eq. (1.1) into matrix,

$$y = X\beta + \epsilon \qquad (1.2)$$

where y is a $N \times 1$ vector of the observed response variable, X is a $N \times (p + 1)$ design matrix including all the independent variables with the first column as one for the intercept (i.e., β_0), $\beta = (\beta_0, \beta_1, \cdots, \beta_p)$ is a $(p + 1)$-vector of regression parameters, and ϵ is the error term. Specifically,

$$y = \begin{pmatrix} y_1 \\ y_2 \\ \vdots \\ y_N \end{pmatrix}, \quad \epsilon = \begin{pmatrix} \epsilon_1 \\ \epsilon_2 \\ \vdots \\ \epsilon_N \end{pmatrix}, \qquad (1.3)$$

$$X = \begin{pmatrix} 1 & x_{11} & \cdots & x_{p1} \\ 1 & x_{12} & \cdots & x_{p2} \\ \vdots & \vdots & \ddots & \vdots \\ 1 & x_{1N} & \cdots & x_{pN} \end{pmatrix} \qquad (1.4)$$

1.2.2 Simple Linear Regression Model

When there is only one variable in X (for example, the variable *Year* in Fig. 1.1), the general multiple linear regression model in Eq. (1.1) is simplified as follows:

$$y_i = \beta_0 + \beta_1 x_{1i} + \epsilon_i \tag{1.5}$$

This equation is a simple linear regression (*SLM*), which is distinguished from Eq. (1.1) with more than one variable in multiple linear regression.

Corresponding to the SLM in Eq. (1.5), the matrix form becomes

$$y = X\beta + \epsilon \tag{1.6}$$

where y is still a $N \times 1$ vector of the observed response variable, X is now a $N \times 2$ design matrix, and $\beta = (\beta_0, \beta_1)$. Specifically,

$$y = \begin{pmatrix} y_1 \\ y_2 \\ \vdots \\ y_N \end{pmatrix}, \quad X = \begin{pmatrix} 1 & x_{11} \\ 1 & x_{12} \\ \vdots & \vdots \\ 1 & x_{1N} \end{pmatrix}, \quad \epsilon = \begin{pmatrix} \epsilon_1 \\ \epsilon_2 \\ \vdots \\ \epsilon_N \end{pmatrix}.$$

1.2.3 The Method of Least Squares Estimation (LSE)

Fundamentally, by estimating the parameter β, we are trying to find values (i.e., estimates) such that the systematic component (i.e., X) explains as much of the variation in the response (i.e., y) as possible. Hence, we are finding parameter values that make the error as small as possible. This is called *least squares estimation*, i.e., to find β so that the sum of the squared errors is as small as possible. Therefore, the least squares estimate (*LSE*) of β, denoted by $\hat{\beta}$, can be obtained by minimizing the sum of squared errors (*SSE*):

$$SSE = \sum_i \epsilon_i^2 = \epsilon'\epsilon = (y - X\beta)'(y - X\beta)$$

$$= y'y - 2\beta X'y + \beta'X'X\beta \tag{1.7}$$

Taking the derivative of the sum of squared errors (SSE) with respect to β and setting to zero leads to

$$X'X\hat{\beta} = X'y \tag{1.8}$$

when $X'X$ is *invertible*, we can have

$$\hat{\beta} = (X'X)^{-1}X'y. \tag{1.9}$$

In this LSE, the parameter estimation $\hat{\beta}$ can be estimated by Eq. (1.9). This estimation is simply calculated with the above analytical equation and is therefore mathematically guaranteed globally with the desired solution.

1.2.4 The Properties of LSE

We list below some results associated with least squares estimation, which are commonly used in the book:

1. *Unbiasedness*: The LSE $\hat{\beta}$ is unbiased since $E(\hat{\beta}) = \beta$ with variance $var(\hat{\beta}) = (X'X)^{-1}\sigma^2$ if $var(\epsilon) = \sigma^2 I$.
2. *The predicted values*: The predicted values are calculated as $\hat{y} = X\hat{\beta} = X(X'X)^{-1}X'y = Hy$, where $H = X(X'X)^{-1}X'$ is called the *hat-matrix* to turn observed y into *hat* $\hat{y}$.
3. *The residuals for diagnostics*: The regression residuals are defined as $\hat{\epsilon} = y - \hat{y} = y - X\hat{\beta} = (I - H)y$, which are the key components for model diagnostics.
4. *Residual sum of squares (RSS)*: The RSS is defined as $RSS = \hat{\epsilon}'\hat{\epsilon} = y'(I - H)'(I - H)y = y'(I - H)y$, which is used to estimate the residual variance and to evaluate the goodness of model fitting.
5. *Variance estimate*: It can be shown that $E(\hat{\epsilon}'\hat{\epsilon}) = \sigma^2(N - p - 1)$ (where $p + 1$ is the number of columns of the design matrix X, i.e., the number of parameters in the linear model). We then *estimate* σ^2 using

$$\hat{\sigma}^2 = \frac{\hat{\epsilon}'\hat{\epsilon}}{N - p - 1} = \frac{RSS}{N - p - 1} \tag{1.10}$$

   where $N - p - 1$ is the *degrees of freedom* of the model.
6. R^2: R^2 is called the *coefficient of determination* or percentage of response variation explained by the systematic component (i.e., X, which is usually used as a goodness of fit measure):

$$R^2 = \frac{\sum(\hat{y}_i - \bar{y})^2}{\sum(y_i - \bar{y})^2} = 1 - \frac{\sum(y_i - \hat{y}_i)^2}{\sum(y_i - \bar{y})^2}. \tag{1.11}$$

R^2 ranges from 0 to 1, with values closer to 1 indicating a better fit of the model.

1.3 Simple Linear Regression: One Independent Variable

1.3.1 Model Fitting

To examine this increase in life expectancy numerically, we can make use of linear regression. The implementation of linear regression in *R* is very straightforward. The *R* function *lm* (i.e., *l*inear *m*odel) will allow you to perform this linear regression.

For example, the simple linear regression (*SLM*) model on *LifeExpectacy* over *Year* can be done as follows:

```
# simple linear regression
slm1 = lm(LifeExpectancy~Year,dWHO)
```

We can print the model summary. The model summary provides information on model fit (overall F-test for significance, R^2, and standard error of the estimate), parameter significance tests, and a summary of residual statistics.

```
summary(slm1)
```

```
## Call:
## lm(formula = LifeExpectancy ~ Year, data = dWHO)
##
## Residuals:
##     Min      1Q  Median      3Q     Max
## -33.803  -6.504   2.852   6.295  21.004
##
## Coefficients:
##                 Estimate Std. Error t value Pr(>|t|)
## (Intercept) -635.87153   75.54552   -8.417   <2e-16 ***
## Year           0.35123    0.03763    9.333   <2e-16 ***
## ---
## Residual standard error: 9.387 on 2926 degrees of freedom
##    (10 observations deleted due to missingness)
## Multiple R-squared:  0.02891,    Adjusted R-squared: 0.02858
## F-statistic: 87.11 on 1 and 2926 DF,  p-value: < 2.2e-16
```

As seen from this model summary, we obtained a significant simple linear regression model as indicated from the significant *F-statistic: 87.11 on 1 and 2926 DF* which produced a *p-value: < 2.2e−16*.

From this simple linear regression model, the estimated slope parameter $\hat{\beta}_1 = 0.3512$ with a standard error of 0.0376 which gave a *p-value < 2e−16*. This indicated that the life expectancy is statistically significantly increasing every year at the rate of 0.3512 per year. However, we should be cautious of this interpretation since the estimated R^2 from this simple linear regression is only 0.028. This means that this significant regression model only explains about 2.8% of the variation in the data. This can be seen from Fig. 1.1 that shows large variations in the data.

The software *R* is an object-oriented language. As such, the model fit is saved as an *R* object. We can access all the additional information that the model produces, such as predicted values and residuals. Using the *R* command *attributes()*, we can obtain a list of the additional information available for the *lm* function:

```
attributes(slm1)
```

```
## $names
##  [1] "coefficients"  "residuals"      "effects"
##  [4] "rank"          "fitted.values"  "assign"
##  [7] "qr"            "df.residual"    "na.action"
## [10] "xlevels"       "call"           "terms"     "model"
## $class
## [1] "lm"
```

For example, to access the LSE parameter estimates, we can call the *coefficients* as follows:

```
slm1$coefficients
```

```
##   (Intercept)            Year
## -635.8715325       0.3512311
```

which would give us the estimated parameters. Interested readers should get familiar with this trick since we will access *R* objects in this book often for data analysis and plotting.

We can use the *R* function *abline* to add the fitted regression model with larger *line width* (i.e., *lwd=3*) and *red* color (i.e., *col="red"*) to the data as seen in Fig. 1.2.

```
### Plot the data
plot(LifeExpectancy~Year,xlab="Year", ylab="Life Expectancy", dWHO)
### Add the fitted regression to the data
abline(slm1, lwd=3, col="red")
```

1.3.2   Model Diagnostics

Before we interpret these results, we have to make sure the model is appropriate. This is done with model diagnostics.

For model diagnostics on residuals, we validate whether or not the residuals are homogeneous and are normally distributed. We can make use of *R* package *car* (i.e., *C*ompanion to *A*pplied *R*egression) for this purpose. Let us load this package first:

```
library(car)
```

The function *residualPlots* in this package can be called to produce a *Tukey* test on residual homogeneity along with the residual plots as shown in Fig. 1.3 with the following *R* code chunk:

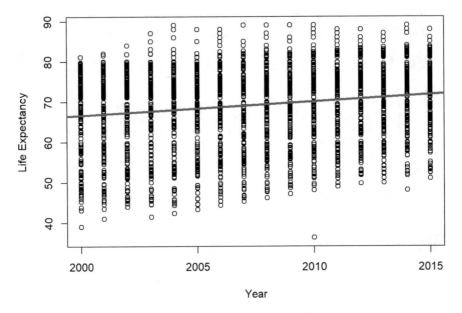

Fig. 1.2 Illustration of model fitting

```
# Residual homogeneity plot and test
residualPlots(slm1)
```

```
##              Test stat Pr(>|Test stat|)
## Year            0.075           0.9402
## Tukey test      0.075           0.9402
```

The *p-values* from the *Tukey* test are both at 0.94, thus indicating that there is no deviation from residual homogeneity. The QQ-plot in Fig. 1.4 also indicated no obvious pattern of deviation from residual normality.

```
# Residual qqnorm plot
qqPlot(slm1)
```

```
## [1] 1133 2298
```

1.4   Multiple Linear Regression: Moderation Analysis

1.4.1   Statistical Interaction/Moderation

Due to the low R^2 from the simple linear regression above, let us examine the life expectancy over each different *Status* to see whether or not country status modifies

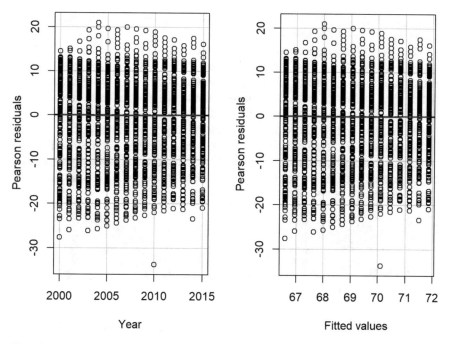

Fig. 1.3 Illustration of residual plots

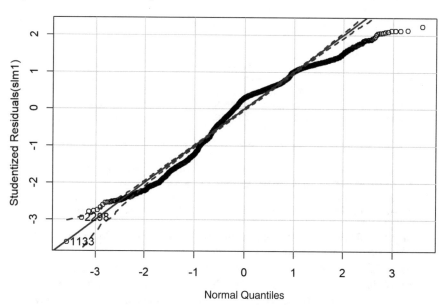

Fig. 1.4 Illustration of residual QQ-plot

life expectancy. In other words, is there a difference in life expectancy between *Developed* and *Developing* countries?

The simple linear regression now becomes multiple linear regression with the added variable *Status*. To implement multiple linear regression in *R*, we can simply add more variables into the *lm* function. In this case, we can simply add *Status* as a statistical interaction term to the above simple linear regression to consider a moderation analysis involving the *Status* variable. Notice that we will use *moderation* and *statistical interaction* interchangeably in this book.

We can include an interaction between *Year* and *Status* to test the effect of *Year* and *Status* in addition to the main effects. This interaction can be simply formulated as the product of these two variables, i.e., $Year * Status$. So, the moderation model (i.e., interaction model) is defined as follows:

```
# multiple linear regression
mlm0 = lm(LifeExpectancy~Year*Status, dWHO)
# summary of the model fitting
summary(mlm0)
```

```
##
## Call:
## lm(formula = LifeExpectancy ~ Year * Status, data = dWHO)
##
## Residuals:
##     Min      1Q  Median      3Q     Max
## -31.725  -5.491   1.146   6.347  22.071
##
## Coefficients:
##                          Estimate Std. Error t value Pr(>|t|)
## (Intercept)            -489.83061  157.58892  -3.108  0.00190 **
## Year                      0.28345    0.07850   3.611  0.00031 ***
## StatusDeveloping       -176.99000  173.48533  -1.020  0.30772
## Year:StatusDeveloping     0.08214    0.08642   0.951  0.34192
## ---
## Signif. codes:  0 '***' 0.001 '**' 0.01 '*' 0.05 '.' 0.1 ' ' 1
##
## Residual standard error: 8.188 on 2924 degrees of freedom
##    (10 observations deleted due to missingness)
## Multiple R-squared:  0.2616, Adjusted R-squared:  0.2608
## F-statistic: 345.3 on 3 and 2924 DF,  p-value: < 2.2e-16
```

This multiple regression is significant and the associated R^2 increased to 0.262 from 0.028 in the previous simple linear regression.

Examining the interaction between *Year* and *Status*, we can see that this interaction is not statistically significant. The slope of the interaction, denoted by *Year:StatusDeveloping*, is estimated as 0.08214 and is not significant (i.e., *t value*= *0.951* and *p-value = 0.3419*). This result indicates that the level of *Status* (i.e., *Developed* vs. *Developing*) does not change or moderate the relationship between *Year* and the *LifeExpectancy* in this dataset.

We can also use the *F-test* for the overall model from the entire ANOVA result including sums of squares and mean squares by using the *anova* function as follows:

```
# anova table for F-test
anova(mlm0)
```

```
## Analysis of Variance Table
##
## Response: LifeExpectancy
##               Df Sum Sq Mean Sq  F value Pr(>F)
## Year           1   7676    7676 114.4851 <2e-16 ***
## Status         1  61715   61715 920.4959 <2e-16 ***
## Year:Status    1     61      61   0.9035 0.3419
## Residuals   2924 196040      67
## ---
## Signif. codes:  0 '***' 0.001 '**' 0.01 '*' 0.05 '.' 0.1 ' ' 1
```

1.4.2 Main-Effect Model

With these results, let us reduce the interaction model to a main-effect model by removing the interaction term (i.e. change *Year:Status* to *Year+Status*) as follows:

```
# multiple linear regression
mlm1 = lm(LifeExpectancy~Year+Status, dWHO)
# summary of the model fitting
summary(mlm1)
```

```
##
## Call:
## lm(formula = LifeExpectancy ~ Year + Status, data = dWHO)
##
## Residuals:
##     Min      1Q  Median      3Q     Max
## -31.690  -5.488   1.222   6.362  22.064
##
## Coefficients:
##                     Estimate Std. Error t value Pr(>|t|)
## (Intercept)       -625.89861   65.89817  -9.498   <2e-16 ***
## Year                 0.35123    0.03283  10.700   <2e-16 ***
## StatusDeveloping   -12.08639    0.39836 -30.340   <2e-16 ***
## ---
## Signif. codes:  0 '***' 0.001 '**' 0.01 '*' 0.05 '.' 0.1 ' ' 1
##
## Residual standard error: 8.188 on 2925 degrees of freedom
##   (10 observations deleted due to missingness)
## Multiple R-squared:  0.2614, Adjusted R-squared:  0.2609
## F-statistic: 517.5 on 2 and 2925 DF,  p-value: < 2.2e-16
```

A better presentation in Table 1.1 of the parameter estimation can be produced by using *R* package *xtable* as follows:

Table 1.1 Summary of the parameter estimation

	Estimate	Std. error	t value	Pr(>|t|)
(Intercept)	−625.8986	65.8982	−9.4980	0
Year	0.3512	0.0328	10.6999	0
StatusDeveloping	−12.0864	0.3984	−30.3402	0

```
# load the library
library(xtable)
# Make the table
knitr::kable(
round(xtable(mlm1),4),caption='Summary of the Parameter Estimation',
   booktabs = TRUE
)
```

This reduced multiple regression is still significant and the associated R^2 dropped to 0.261 in the main-effect model *mlm1* from 0.262 in the interaction model *mlm0*.

From this multiple linear regression, the estimated parameter for *Status* associated with *Developing* countries is reduced by 12 years (i.e., $\hat{\beta}_2 = -12.086$) after adjustment for *Years*. More specifically, the life expectancy in *Developing* countries is about 12 years shorter on average than that of "Developed" countries, which is a statistically significant reduction.

To illustrate this moderation effect graphically, we can make use of the following *R* code chunk to produce Fig. 1.5. In this figure, we use + (i.e., *pch* (plot *ch*aracter) = 3), for data in *Developed* countries and × (i.e., *pch* = 4) for *Developing* countries. The blue dashed line is for the fitted regression for the *Developed* countries and red dashed line for *Developing* countries.

```
# Plot the data
plot(LifeExpectancy~Year,xlab="Year",las=1,pch=3:4,col=c("blue",
    "red"), ylab="Life Expectancy", dWHO)

# Add the fitted lines to the plot
abline(mlm1$coef[1],mlm1$coef[2],lwd=3,lty=8, col="blue")
abline(mlm1$coef[1]+mlm1$coef[3],mlm1$coef[2],lwd=3, lty=4,
    col="red")
```

1.5 Multiple Linear Regression: Stepwise Model Selection

1.5.1 R *Stepwise Model Selection Algorithm:* Step

We can further explore multiple linear regression by using all the variables in the data to investigate which factors or covariates will be significantly related to *LifeExpectancy*. This is called *data-driven modeling*. For this purpose, we can call *lm* for multiple linear regression and explore another *R* function *step* to choose a

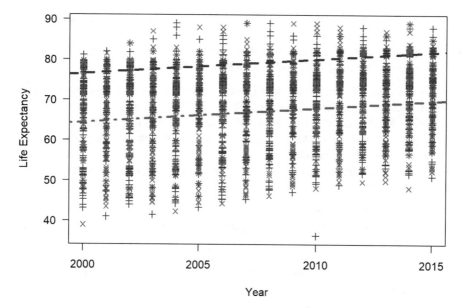

Fig. 1.5 Illustration of moderation

model by Akaike Information Criterion (AIC) in a *Stepwise Algorithm*. Note that *AIC* is a measure of model goodness of fit which penalizes the model for having more variables (because you might add more variables to increase the R^2, but the new variables do not add to the overall model fit at all), where the smaller the AIC, the better the model fit.

The algorithm is to take the object from *lm*, then *step*-wisely *add1* and *drop1* repeatedly based on the *AIC* value from each model. The usage of this algorithm is as follows:

```
step(object, scope, scale = 0,
    direction = c("both", "backward", "forward"),
    trace = 1, keep = NULL, steps = 1000, k = 2, ...)
```

where

- *object* is an object representing a model of an appropriate class, such as in *lm* and *glm*, which is used as the initial model in the stepwise search,
- *scope* defines the range of models examined in the stepwise search,
- *scale* is used in the definition of the AIC statistic for selecting the models,
- *direction* defines the mode of stepwise search which can be one of *both*, *backward*, or *forward*, with a default of "both,"
- *trace* is the control information to be printed during the running of step,
- *keep* is a filter function whose input is a fitted model object and the associated AIC statistic, and whose output is arbitrary,

- *steps* is the maximum number of steps to be considered with default of 1000 (essentially as many as required),
- *k* is the multiple of the number of degrees of freedom used for the penalty where k = 2 for the genuine AIC and k = log(N) for BIC or SBC.

To use *step*, all the models fitted in *step* must be to the same dataset. This may create a problem if there are missing values for different variables in the data. In this situation, the dataset in each *step*-wise selected model would be different. Therefore to implement *step* for model selection, it is recommended to remove the missing values first before model selection.

With this explanation, the typical stepwise model selection is implemented as in following steps:

1. remove all the missing values,
2. re-run the multiple linear regression *lm* for all variables (without *Country* since otherwise the *lm* would output too much for each country),
3. print the model output for reference,
4. implement the *step* algorithm with selection of *both* for *forward-and-backward* selection,
5. print the final model from this *step*-wise algorithm.

1.5.2   Data Analysis with step Algorithm

The following *R* code chunk is to implement the above five steps for real data analysis:

```
# 1. Remove all the missing values using R function *complete.cases*
dt = dWHO[complete.cases(dWHO),]
# check the data dimmension before/after removing missing values
dim(dWHO);dim(dt)
```

```
## [1]  2938     22
```

```
## [1]  1649     22
```

```
# 2. Multiple linear regression with all variables, but
  remove *Country*
mlm2 = lm(LifeExpectancy~., data=dt[,-1])

# 3. Print the summary of model fit
summary(mlm2)
```

```
##
## Call:
## lm(formula = LifeExpectancy ~ ., data = dt[, -1])
##
## Residuals:
```

```
##       Min        1Q   Median       3Q      Max
## -16.7681   -2.1427   0.0273   2.1776  12.4242
##
## Coefficients:
##                          Estimate Std. Error t value Pr(>|t|)
## (Intercept)             3.090e+02  4.621e+01   6.687 3.12e-11 ***
## Year                   -1.272e-01  2.308e-02  -5.510 4.18e-08 ***
## StatusDeveloping       -8.865e-01  3.353e-01  -2.644  0.00827 **
## AdultMortality         -1.621e-02  9.441e-04 -17.171  < 2e-16 ***
## InfantDeaths            8.873e-02  1.059e-02   8.376  < 2e-16 ***
## Alcohol                -1.313e-02  3.366e-02  -3.901 9.95e-05 ***
## PctExpenditure          3.026e-04  1.789e-04   1.691  0.09096 .
## HepatitisB             -3.258e-03  4.449e-03  -0.732  0.46413
## Measles                -1.033e-05  1.070e-05  -0.966  0.33439
## BMI                     3.183e-02  5.955e-03   5.345 1.03e-07 ***
## UnderFiveDeaths        -6.662e-02  7.673e-03  -8.682  < 2e-16 ***
## Polio                   5.797e-03  5.121e-03   1.132  0.25776
## TotalExpenditure        9.220e-02  4.042e-02   2.281  0.02268 *
## Diphtheria              1.403e-02  5.877e-03   2.387  0.01712 *
## HIVAIDS                -4.481e-01  1.780e-02 -25.174  < 2e-16 ***
## GDP                     2.451e-05  2.826e-05   0.867  0.38594
## Population             -6.085e-10  1.733e-09  -0.351  0.72558
## Thin1to19              -5.815e-03  5.254e-02  -0.111  0.91189
## Thin5to9               -5.010e-02  5.185e-02  -0.966  0.33412
## Income                  1.045e+01  8.327e-01  12.549  < 2e-16 ***
## Schooling               8.949e-01  5.910e-02  15.142  < 2e-16 ***
## ---
## Signif. codes: 0 '***' 0.001'**'0.01 '*' 0.05 '.' 0.1 ' ' 1
##
## Residual standard error: 3.556 on 1628 degrees of freedom
## Multiple R-squared:  0.8386, Adjusted R-squared:  0.8366
## F-statistic: 422.9 on 20 and 1628 DF,  p-value: < 2.2e-16
```

```
# 4. Call *step* for model selection
mlm3 = step(mlm2,direction="both")
```

```
## Start:  AIC=4204.82
## LifeExpectancy ~ Year + Status + AdultMortality + InfantDeaths +
##     Alcohol + PctExpenditure + HepatitisB + Measles + BMI +
##     UnderFiveDeaths + Polio + TotalExpenditure + Diphtheria +
##     HIVAIDS + GDP + Population + Thin1to19 + Thin5to9 +
##     Income + Schooling
##
##                    Df Sum of Sq   RSS    AIC
## - Thin1to19         1       0.2 20586 4202.8
## - Population        1       1.6 20588 4202.9
## - HepatitisB        1       6.8 20593 4203.4
## - GDP               1       9.5 20596 4203.6
## - Measles           1      11.8 20598 4203.8
## - Thin5to9          1      11.8 20598 4203.8
## - Polio             1      16.2 20603 4204.1
## <none>                           20586 4204.8
## - PctExpenditure    1      36.2 20622 4205.7
```

```
## - TotalExpenditure  1        65.8 20652 4208.1
## - Diphtheria        1        72.0 20658 4208.6
## - Status            1        88.4 20675 4209.9
## - Alcohol           1       192.5 20779 4218.2
## - BMI               1       361.2 20948 4231.5
## - Year              1       383.8 20970 4233.3
## - InfantDeaths      1       887.1 21473 4272.4
## - UnderFiveDeaths   1       953.2 21540 4277.5
## - Income            1      1991.4 22578 4355.1
## - Schooling         1      2899.2 23486 4420.1
## - AdultMortality    1      3728.2 24314 4477.3
## - HIVAIDS           1      8013.7 28600 4745.0
##
## Step:  AIC=4202.84
## LifeExpectancy ~ Year + Status + AdultMortality + InfantDeaths +
##     Alcohol + PctExpenditure + HepatitisB + Measles + BMI +
##     UnderFiveDeaths + Polio + TotalExpenditure + Diphtheria +
##     HIVAIDS + GDP + Population + Thin5to9 + Income + Schooling
##
##                     Df Sum of Sq   RSS    AIC
## - Population        1         1.6 20588 4201.0
## - HepatitisB        1         6.8 20593 4201.4
## - GDP               1         9.5 20596 4201.6
## - Measles           1        11.8 20598 4201.8
## - Polio             1        16.1 20603 4202.1
## <none>                            20586 4202.8
## - PctExpenditure    1        36.2 20623 4203.7
## + Thin1to19         1         0.2 20586 4204.8
## - Thin5to9          1        54.6 20641 4205.2
## - TotalExpenditure  1        65.7 20652 4206.1
## - Diphtheria        1        72.3 20659 4206.6
## - Status            1        88.3 20675 4207.9
## - Alcohol           1       192.6 20779 4216.2
## - BMI               1       363.0 20949 4229.7
## - Year              1       384.4 20971 4231.3
## - InfantDeaths      1       888.0 21474 4270.5
## - UnderFiveDeaths   1       954.9 21541 4275.6
## - Income            1      1996.0 22582 4353.4
## - Schooling         1      2912.2 23499 4419.0
## - AdultMortality    1      3730.6 24317 4475.5
## - HIVAIDS           1      8016.8 28603 4743.2
##
## Step:  AIC=4200.96
## LifeExpectancy ~ Year + Status + AdultMortality + InfantDeaths +
##     Alcohol + PctExpenditure + HepatitisB + Measles + BMI +
##     UnderFiveDeaths + Polio + TotalExpenditure + Diphtheria +
##     HIVAIDS + GDP + Thin5to9 + Income + Schooling
##
##                     Df Sum of Sq   RSS    AIC
## - HepatitisB        1         6.7 20595 4199.5
## - GDP               1         9.6 20598 4199.7
## - Measles           1        11.1 20599 4199.9
## - Polio             1        16.0 20604 4200.2
## <none>                            20588 4201.0
```

```
## - PctExpenditure     1       35.9 20624 4201.8
## + Population         1        1.6 20586 4202.8
## + Thin1to19          1        0.2 20588 4202.9
## - Thin5to9           1       54.6 20643 4203.3
## - TotalExpenditure   1       65.8 20654 4204.2
## - Diphtheria         1       71.8 20660 4204.7
## - Status             1       88.5 20677 4206.0
## - Alcohol            1      192.6 20781 4214.3
## - BMI                1      361.8 20950 4227.7
## - Year               1      384.6 20973 4229.5
## - InfantDeaths       1      924.1 21512 4271.4
## - UnderFiveDeaths    1      971.4 21559 4275.0
## - Income             1     1998.0 22586 4351.7
## - Schooling          1     2916.2 23504 4417.4
## - AdultMortality     1     3741.5 24330 4474.3
## - HIVAIDS            1     8015.3 28603 4741.2
##
## Step:  AIC=4199.5
## LifeExpectancy ~ Year + Status + AdultMortality + InfantDeaths +
##      Alcohol + PctExpenditure + Measles + BMI + UnderFiveDeaths +
##      Polio + TotalExpenditure + Diphtheria + HIVAIDS + GDP +
##      Thin5to9 + Income + Schooling
##
##                     Df Sum of Sq  RSS    AIC
## - GDP                1        9.4 20604 4198.3
## - Measles            1       10.9 20606 4198.4
## - Polio              1       13.1 20608 4198.5
## <none>                          20595 4199.5
## - PctExpenditure     1       37.4 20632 4200.5
## + HepatitisB         1        6.7 20588 4201.0
## + Population         1        1.4 20593 4201.4
## + Thin1to19          1        0.2 20594 4201.5
## - Thin5to9           1       56.3 20651 4202.0
## - TotalExpenditure   1       64.6 20659 4202.7
## - Diphtheria         1       66.5 20661 4202.8
## - Status             1       85.2 20680 4204.3
## - Alcohol            1      190.8 20786 4212.7
## - BMI                1      357.6 20952 4225.9
## - Year               1      405.5 21000 4229.7
## - InfantDeaths       1      926.8 21522 4270.1
## - UnderFiveDeaths    1      971.9 21567 4273.5
## - Income             1     2011.0 22606 4351.1
## - Schooling          1     2917.3 23512 4416.0
## - AdultMortality     1     3750.9 24346 4473.4
## - HIVAIDS            1     8008.7 28603 4739.2
##
## Step:  AIC=4198.25
## LifeExpectancy ~ Year + Status + AdultMortality + InfantDeaths +
##      Alcohol + PctExpenditure + Measles + BMI + UnderFiveDeaths +
##      Polio + TotalExpenditure + Diphtheria + HIVAIDS + Thin5to9 +
##      Income + Schooling
##
##                     Df Sum of Sq  RSS    AIC
## - Measles            1       10.9 20615 4197.1
```

```
## - Polio              1        14.0 20618 4197.4
## <none>                            20604 4198.3
## + GDP                1         9.4 20595 4199.5
## + HepatitisB         1         6.5 20598 4199.7
## + Population         1         1.5 20603 4200.1
## + Thin1to19          1         0.2 20604 4200.2
## - Thin5to9           1        57.4 20662 4200.8
## - TotalExpenditure   1        62.6 20667 4201.3
## - Diphtheria         1        66.2 20670 4201.5
## - Status             1        89.5 20694 4203.4
## - Alcohol            1       188.7 20793 4211.3
## - BMI                1       355.9 20960 4224.5
## - Year               1       397.1 21001 4227.7
## - PctExpenditure     1       748.3 21352 4255.1
## - InfantDeaths       1       928.3 21532 4268.9
## - UnderFiveDeaths    1       973.3 21577 4272.4
## - Income             1      2033.1 22637 4351.4
## - Schooling          1      2958.8 23563 4417.5
## - AdultMortality     1      3750.3 24354 4472.0
## - HIVAIDS            1      8004.5 28609 4737.5
##
## Step:  AIC=4197.12
## LifeExpectancy ~ Year + Status + AdultMortality + InfantDeaths +
##     Alcohol + PctExpenditure + BMI + UnderFiveDeaths + Polio +
##     TotalExpenditure + Diphtheria + HIVAIDS + Thin5to9 + Income +
##     Schooling
##
##                     Df Sum of Sq   RSS    AIC
## - Polio              1        13.9 20629 4196.2
## <none>                            20615 4197.1
## + Measles            1        10.9 20604 4198.3
## + GDP                1         9.4 20606 4198.4
## + HepatitisB         1         6.4 20609 4198.6
## + Population         1         0.9 20614 4199.1
## + Thin1to19          1         0.2 20615 4199.1
## - Thin5to9           1        51.5 20667 4199.2
## - TotalExpenditure   1        64.8 20680 4200.3
## - Diphtheria         1        66.4 20681 4200.4
## - Status             1        91.2 20706 4202.4
## - Alcohol            1       192.0 20807 4210.4
## - BMI                1       371.8 20987 4224.6
## - Year               1       392.1 21007 4226.2
## - PctExpenditure     1       750.7 21366 4254.1
## - InfantDeaths       1       939.1 21554 4268.6
## - UnderFiveDeaths    1       973.3 21588 4271.2
## - Income             1      2031.8 22647 4350.1
## - Schooling          1      2969.8 23585 4417.0
## - AdultMortality     1      3747.8 24363 4470.6
## - HIVAIDS            1      8017.7 28633 4736.9
##
## Step:  AIC=4196.23
## LifeExpectancy ~ Year + Status + AdultMortality + InfantDeaths +
##     Alcohol + PctExpenditure + BMI + UnderFiveDeaths +
##     TotalExpenditure + Diphtheria + HIVAIDS + Thin5to9 +
```

```
##       Income + Schooling
##
##                     Df Sum of Sq   RSS    AIC
## <none>                            20629 4196.2
## + Polio              1      13.9 20615 4197.1
## + Measles            1      10.8 20618 4197.4
## + GDP                1      10.2 20619 4197.4
## + HepatitisB         1       3.5 20625 4198.0
## + Population         1       0.8 20628 4198.2
## + Thin1to19          1       0.0 20629 4198.2
## - Thin5to9           1      50.1 20679 4198.2
## - TotalExpenditure   1      65.8 20695 4199.5
## - Status             1      91.8 20721 4201.6
## - Diphtheria         1     143.2 20772 4205.6
## - Alcohol            1     189.7 20819 4209.3
## - BMI                1     371.1 21000 4223.6
## - Year               1     400.5 21029 4225.9
## - PctExpenditure     1     746.5 21375 4252.9
## - InfantDeaths       1     951.7 21581 4268.6
## - UnderFiveDeaths    1     987.8 21617 4271.4
## - Income             1    2033.2 22662 4349.2
## - Schooling          1    3030.5 23659 4420.3
## - AdultMortality     1    3767.7 24397 4470.9
## - HIVAIDS            1    8019.0 28648 4735.7
```

```
# 5. Print the model selection
summary(mlm3)
```

```
##
## Call:
## lm(formula = LifeExpectancy ~ Year + Status + AdultMortality +
##       InfantDeaths + Alcohol + PctExpenditure + BMI + UnderFiveDeaths +
##       TotalExpenditure + Diphtheria + HIVAIDS + Thin5to9 + Income +
##       Schooling, data = dt[, -1])
##
## Residuals:
##      Min       1Q   Median       3Q      Max
## -16.7779  -2.1865   0.0023   2.2038  12.4209
##
## Coefficients:
##                    Estimate Std. Error t value Pr(>|t|)
## (Intercept)       3.101e+02  4.540e+01   6.831 1.19e-11 ***
## Year             -1.277e-01  2.268e-02  -5.632 2.09e-08 ***
## StatusDeveloping -8.975e-01  3.329e-01  -2.696 0.007089 **
## AdultMortality   -1.626e-02  9.413e-04 -17.275  < 2e-16 ***
## InfantDeaths      8.608e-02  9.914e-03   8.682  < 2e-16 ***
## Alcohol          -1.299e-01  3.351e-02  -3.877 0.000110 ***
## PctExpenditure    4.523e-04  5.882e-05   7.690 2.53e-14 ***
## BMI               3.199e-02  5.901e-03   5.421 6.80e-08 ***
## UnderFiveDeaths  -6.516e-02  7.366e-03  -8.846  < 2e-16 ***
## TotalExpenditure  9.201e-02  4.029e-02   2.284 0.022516 *
## Diphtheria        1.509e-02  4.481e-03   3.368 0.000776 ***
## HIVAIDS          -4.478e-01  1.777e-02 -25.203  < 2e-16 ***
## Thin5to9         -5.212e-02  2.616e-02  -1.992 0.046497 *
## Income            1.052e+01  8.292e-01  12.690  < 2e-16 ***
## Schooling         9.047e-01  5.839e-02  15.493  < 2e-16 ***
```

```
## ---
## Residual standard error: 3.553 on 1634 degrees of freedom
## Multiple R-squared:  0.8382, Adjusted R-squared:  0.8369
## F-statistic: 604.8 on 14 and 1634 DF,  p-value: < 2.2e-16
```

1.5.3 Discussions on Stepwise Regression

As seen from the sizable output, we started with the dataframe *dWHO* with 2938 observations and 22 variables. After removing the missing values, the new dataframe *dt* has 1649 observations and 22 variables.

We then took this new dataframe *dt* for a multiple regression model using outcome variable *LifeExpectancy*and all the independent variables of *Year + Status + AdultMortality + InfantDeaths + Alcohol + PctExpenditure + HepatitisB + Measles + BMI + UnderFiveDeaths + TotalExpenditure + Diphtheria + HIVAIDS + GDP + Population + Thin1to19 + Thin5to9 + Income + Schooling* to build *mlm2* (i.e., model 2). With this model, the *F-statistic = 422.9 on 20 and 1628 DF*, which yielded the *p-value: < 2.2e-16* with a statistically significant regression model. With this model the *Multiple R-squared = 0.8386* and *Adjusted R-squared = 0.8366*.

In this *mlm2*, the *LifeExpectancy* is statistically significantly (at 5% significant level) related to the independent variables of *Year + Status + AdultMortality + InfantDeaths + Alcohol + BMI + UnderFiveDeaths + TotalExpenditure + Diphtheria + HIVAIDS + Income + Schooling*, but not statistically significantly related to *PctExpenditure + HepatitisB + Measles + Polio + GDP + Population + Thin1to19 + Thin5to9*.

Based on this *mlm2*, we proceeded to use *step*-wise model selection with *both* (i.e., both *forward* and *backward* model selection) using the model selection criterion *AIC* as follows:

- *Step 0*: As the initial step with *mlm2*, the *AIC = 4204.82*,
- *Step 1*: In this step, the most non-significant variable *Thin1to19* was selected out, which reduced the *AIC* to *4202.84* from *4204.82*,
- *Step 2*: In this step, the next non-significant variable *Population* was selected out, which reduced the *AIC* to *4200.96* from *4202.84*,
- *Step 3*: In this step, the next non-significant variable *HepatitisB* was selected out, which reduced the *AIC* to *4199.5* from *4200.96*,
- *Step 4*: In this step, the next non-significant variable *GDP* was selected out, which reduced the *AIC* to *4198.25* from *4199.5*,
- *Step 5*: In this step, the next non-significant variable *Measles* was selected out, which reduced the *AIC* to *4197.12* from *4198.25*,
- *Step 6*: In this step, the next non-significant variable *Polio* was selected out, which reduced the *AIC* to *4196.23* from *4197.12*.

The model selection stopped at *Step 6*. With this model selection process, the selected model *mlm3*, is statistically significant as seen from the *F-statistic: 604.8* *on 14 and 1634 DF, p-value: <2.2e−16.* With this model, the R^2 increased to 0.8382 in *mlm3* from 0.776 in *mlm2* and 0.261 in *mlm1*.

With a series of steps in stepwise model selection, the final model selected is in *mlm3* (i.e., Model 3), which indicated that the life expectancy is statistically significantly related to a subset of variables, including *Year, Status, AdultMortality, InfantDeaths, Alcohol, PctExpenditure, BMI, UnderFiveDeaths, TotalExpenditure, Diphtheria, HIVAIDS, Thin5to9, Income, Schooling.*

Further explorations should be done to see whether or not any of these factors interact with one another and are associated with life expectancy. We leave this exploration as exercises for interested readers.

1.6 Regression Model Diagnostics: Checking Assumptions

The basic assumptions for linear regression are residual normality, homogeneity (i.e., constant variance or homoscedasticity), and independence. If any of these assumptions are violated, the statistical conclusions might be wrong in parameter estimate, standard errors, *t*-values, and *p*-values.

Assumptions should be checked for each data analysis. We illustrate the model diagnostics using the final model *mlm3* from the above *step*-wise model selection.

1.6.1 Normality

For residual normality, we would extract the regression residuals and then examine the distribution of the residuals using *hist* as shown in Fig. 1.6 with the following R code chunk. It can be seen that the distribution is symmetrical with center at zero.

```
# Histogram plot to show normality
hist(mlm3$residuals, main="Residual Distribution", xlab="Residuals from mlm3")
```

Figure 1.6 shows a well-behaved one-mode symmetrical distribution with an approximate *bell* shape.

We can also use the *QQ-plot* to plot the quantiles of the residuals to the theoretical normal quantiles to see whether or not there is a one-to-one relationship. In the QQ-plot, the *y*-axis is for the quantiles from the observed residuals and the *x*-axis is for the corresponding points of a normal distribution. The observed data (i.e., residuals) points in quantiles are plotted to the theoretical normal quantiles by black circles. The solid line represents the data conforming perfectly to the normal distribution. Therefore, the closer the observed data in black circles are to the solid line, the more closely the data conforms to the normal distribution.

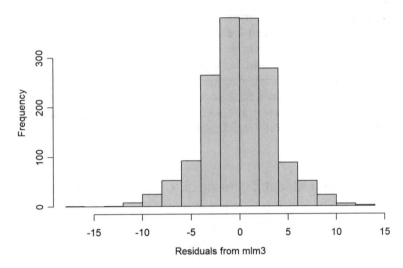

Fig. 1.6 Residual histogram plot

There are two ways to show the QQ-plot. The first way as shown in Fig. 1.7 is to use the default *stat* package and extract the residuals from the *lm* object as shown below:

```
# QQ-plot with residuals extracted from *mlm3*
qqnorm(mlm3$resid)
qqline(mlm3$resid)
```

The second way as shown in Fig. 1.8 is to use *car* package which will show the associated 95% confidence band for the line. When data points fall within the 95% confidence band, they are considered to conform to the normal distribution. We call the *qqPlot* to make this QQ-plot as follows:

```
# QQ-plot with *car* package
qqPlot(mlm3, distribution="norm")
```

```
## 1891 2305
## 1111 1339
```

From the histogram in Fig. 1.6, the distribution of the residuals is symmetric and seems normally distributed. From the QQ-plot in Fig. 1.8, we can see that most of the quantile points are close to the one-to-one line and within the 95% confidence bands except a few points at both ends.

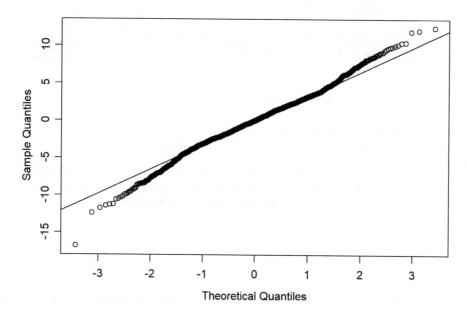

Fig. 1.7 Residual QQ-plot with default *stat* package

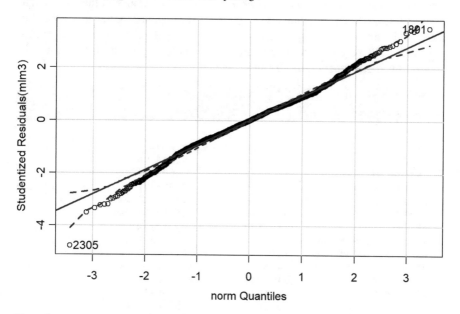

Fig. 1.8 Residual QQ-plot from *car* package

If a statistical test is needed, we can use the *shapiro.test* to perform the Shapiro–Wilk test of normality as follows:

```
## Shapiro test for normality
shapiro.test(mlm3$residuals)
```

```
##
##  Shapiro-Wilk normality test
##
## data:  mlm3$residuals
## W = 0.99168, p-value = 4.694e-08
```

The *p-value* associated with the Shapiro–Wilk test is quite small, indicating that we reject the null hypothesis that the residuals from model *mlm3* are normally distributed. This is mostly because of the outliers shown in Fig. 1.8.

1.6.2 Homogeneity

The residual plot is typically used for homogeneity test (i.e., constant variance). In the *car* package, this is implemented in the function *residualPlots*. This function plots the residuals versus each variable and versus fitted values in the fitted model *mlm3*. In addition, this function gives a curvature test for each of the plots by adding a quadratic term and testing the quadratic term to be zero. For a linear regression model, this is Tukey's test for non-additivity when plotting against fitted values.

Let us call this function for the model *mlm3*:

```
# Residual plot
residualPlots(mlm3,layout=c(3,5))
```

```
##                  Test stat Pr(>|Test stat|)
## Year                0.6148         0.538794
## AdultMortality     -3.1237         0.001817 **
## InfantDeaths        1.7769         0.075772 .
## Alcohol            -4.8478        1.367e-06 ***
## PctExpenditure     -4.4395        9.622e-06 ***
## BMI                 0.3622         0.717272
## UnderFiveDeaths     2.4254         0.015399 *
## TotalExpenditure   -1.5072         0.131947
## Diphtheria          4.6854        3.026e-06 ***
## HIVAIDS            11.3843        < 2.2e-16 ***
## Thin5to9            9.5558        < 2.2e-16 ***
## Income             18.8826        < 2.2e-16 ***
## Schooling          -1.8697         0.061707 .
## Tukey test          6.8709        6.380e-12 ***
## ---
## Signif. codes:  0 '***' 0.001 '**' 0.01 '*' 0.05 '.' 0.1 ' ' 1
```

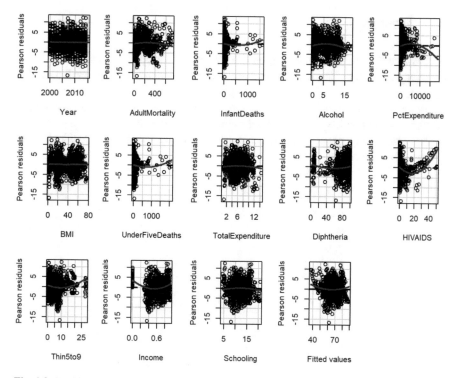

Fig. 1.9 Residual plot to test homogeneity

As seen from the output, no quadratic terms are needed for *Year, InfantDeaths, BMI, TotalExpenditure, Schooling*, but needed for the rest of the variables as shown in Fig. 1.9. This means that quadratic terms should be added to these variables in the multiple linear regression to model the life expectancy. We leave this to interested readers as an exercise.

The Tukey's test for homogeneity is statistically significant with *p-value* = *6.380e-12* indicating that curvature exists in the residuals as seen in the last plot from Fig. 1.9. This is due to those outliers, which is not surprising in real data analysis.

1.6.3   Independence

Another assumption in regression is that the residuals should be independent. There is no statistical test to be used for this purpose. We will see that multi-level modeling is one of the methods that incorporate dependent structures, which is the purpose for this book.

1.7   Exercises

1. Explore the multiple linear regression, i.e., *mlm2* for this data by adding (more) interactions that you may think are practically important in predicting life expectancy.
2. Re-evaluate the residual diagnostics for your selected model to see whether or not you can identify a better model to model life expectancy.
3. Refit *mlm3* adding quadratic terms into *mlm3* and re-evaluate the residuals plots using *residualPlots*.

Chapter 2
Introduction to Multi-Level Modeling

With the overview of classical linear regression and its model diagnostics in Chap. 1, we now have a good understanding of linear regression modeling and the associated assumptions that make a classical regression model valid. If data are clustered (i.e., in multi-level data), the independent assumption in classical linear regression is violated. In this situation, we need to have new regression modeling techniques. Therefore, this chapter is to introduce multi-level modeling with implementation in *R* to analyze multi-level data structures.

2.1 Multi-Level Data Structure

2.1.1 Sampling

All studies face time and budgetary restrictions. As such, sampling an entire population is generally unfeasible. Instead, most studies sample a subset of a larger population because carefully selected samples can provide highly accurate measures of a larger population. Therefore, sampling is an efficient means of examining one or more variables of interest within a target population.

In social interventions and public health research, sampling refers to selecting a subset of individuals from a target population (i.e., population of interest). Characteristics of the target population are then estimated based on the subsample. For example, we would like to estimate the properties of a target population, such as education level, income, and other outcomes. From these outcomes, statistical analyses of the sample allow us to make inferences about the target population's outcomes. These inferences are grounded in both statistical and probability theory.

We review three commonly used sampling techniques: *simple random sampling*, *stratified sampling*, and *cluster sampling*. More comprehensive presentation of sampling techniques in the social sciences and public health can be found in other sampling books, such as Levy and Lemeshow (2008) and Thompson (2012).

2.1.1.1 Simple Random Sampling

Simple random sampling is very commonly used in randomized controlled trials (RCTs). It is the simplest sampling scheme in which a subset of individuals is chosen randomly and entirely by chance from a given population. In simple random sampling, each individual has the same probability of being chosen during the sampling process. Conceptually, simple random sampling requires complete sampling from a given population, which may not be feasible or necessary. More efficient approaches may be utilized if useful information about the population exists. Simple random sampling is the base for most regression models.

2.1.1.2 Stratified Sampling

Stratified sampling is to partition the population into subgroups (i.e., *strata*) based on a factor that may influence the variable of interest. For example, say you want to study the income of individuals who completed a substance use treatment. In your sample, you notice two distinct groups, those who utilized social support groups outside of treatment and those who did not. Collectively, these two groups are called *strata*, and each individual group is called a *stratum*. We then obtain a simple random sample from each stratum. Stratified sampling works best when a heterogeneous population is split into fairly homogeneous *strata*. Under these conditions, stratification generally produces more precise estimates of the population than a simple random sample.

2.1.1.3 Cluster Sampling

Cluster sampling is very different from stratified sampling. In cluster sampling, a population is first split into subgroups (i.e., *clusters*). Second, a simple random sample of clusters is selected. Third, of the sampled clusters, units (e.g., individuals) within the clusters are all taken. For example, say you want to examine the political opinions of all people in the State of North Carolina, USA, which can be too time-consuming. You could cluster North Carolina by counties (i.e., *county* as *cluster*). Next, a random sample of counties (i.e., *clusters*) would be selected, and then all people within the selected clusters (i.e., counties) would be surveyed.

Importantly, and unlike stratified sampling, each cluster can be heterogeneous and data within each cluster could be highly correlated. Statistical analysis for cluster sampling is more complicated than stratified sampling, which is the main focus for multi-level modeling.

2.1.2   *Multi-Level Data or Nested Data*

Cluster sampling produces data with a multi-level data structure, typically referred to as a nested data structure. This type of structure allows for rigorous data analysis without being held to a typical randomized clinical trial design that uses simple random sampling. For example, perhaps you want to examine the impact of a new teaching intervention on student performance. With a typical RCT design that uses simple random sampling, randomization happens at the individual level, and you would randomly select and assign students to an intervention or control condition. Such a design might be impractical logistically and financially.

Alternatively, we could employ a cluster-randomized design and randomize at the cluster level. In the above example, we could examine the impact of a new teaching intervention on student performance in a state using a multi-level or nested design. In this design, *classrooms* act as clusters based on the assigned teachers. Each cluster (i.e., classroom) will be randomly assigned to act as either a treatment or control condition. Hence, if a specific classroom within a school was assigned to the treatment condition, all students in that classroom would receive the new teaching intervention.

Due to clustering, student outcomes within the classrooms are dependent on each other since students within a classroom are influenced by the same teacher in this classroom. Therefore, the independence assumption from a regression/ANOVA model is violated. Further, over and above the impact of the new intervention, it is reasonable to assume that classroom-level factors (such as teachers' education level, etc.) impact student performance. Such impacts influence the correlations in student's test scores among individuals in that class. We refer to this type of data as multi-level data or data with a nested structure.

In its most basic form, multi-level data with two levels would be in the format as follows:

- *Level 1*: individual students who are nested within classroom;
- *Level 2*: classes randomly assigned to intervention/control conditions.

This 2-level nested structure can be easily expanded to include a 3-level (such as classes nested within schools), a 4-level (such as schools nested within school districts), a 5-level (such as districts nested within states), or even higher-level nesting.

2.2   Intra-Class Correlation

As the first concept to quantify the correlation existed in the nested data, the intra-class correlation (*ICC*) is always used to measure this correlation among individuals' outcome measures within the cluster or nested structure.

This *ICC* is a very important measure in nested data analysis. We explain the theoretical background on how to estimate this *ICC* but will illustrate the practical calculations later with *R* package *nlme* since the relevant calculations are done in the model estimation, and we only need to extract those values.

2.2.1   Definition

The intra-class correlation (*ICC*), denoted as ρ_I, is defined as the proportion of variation in the outcome's between-cluster variance versus the total variation present in the data. It ranges from 0 (no variance among clusters) to 1 (variance among clusters but no within-cluster variance). This ρ_I can also be conceptualized as the correlation for the dependent measure for two individuals randomly selected from the same cluster.

It can be expressed as

$$\rho_I = \frac{\tau^2}{\tau^2 + \sigma^2}, \tag{2.1}$$

where τ^2 denotes the population of between-cluster variance and σ^2 indicates the population of within-cluster variance. Higher values of ρ_I indicate that the total variation in the outcome measure is associated with cluster membership; that is, a relatively strong relationship exists among clustered individuals in comparison to individuals between clusters. This means individuals within the same cluster (e.g., classroom) are more alike on a measured variable than they are like individuals in other clusters. This suggests a within-cluster effect is present.

2.2.2   Why ICC *Is the Correlation for Individuals Within Clusters*

The first question that may come up from any interested reader could be why the definition in Eq. (2.1) is the correlation between any two individuals within a cluster.

It seems that from the definition in Eq. (2.1), the *ICC*, ρ_I, which is defined as the proportion of between-cluster variance of τ^2 to the total variances of $\tau^2 + \sigma^2$ (i.e., the between-cluster and within-cluster together), is all about variances from both hierarchical levels. It appears to have nothing to do with the *within-cluster correlation* between any two individuals. However, this *ICC* is intrinsically the correlation between any two individuals within any clusters.

Let us show this fact from the 2-level random-intercept model, which will be discussed in detail in Sect. 2.4. A general 2-level random-intercept MLM model is denoted by

$$y_{ij} = \beta_{0j} + \beta_{1j}x_{1ij} + \epsilon_{ij} = \beta_0 + U_{0j} + \beta_1 x_{1ij} + \epsilon_{ij}, \tag{2.2}$$

where j denotes clusters from $j = 1, \cdots, J$ (i.e., J is the total number of clusters), and i ($i = 1, \cdots, n_j$, and n_j is the number of observations in the jth cluster) denotes the ith observation from the jth cluster. In this formulation, the total number of observations is $N = \sum_{j=1}^{J} n_j$.

Since this is a random-intercept model, the between-cluster variation is described by the error terms on between-cluster intercepts only (not in the slopes), i.e., $U_{0j} \sim N(0, \tau^2)$. The within-cluster errors, ϵ_{ij}, are assumed to be independently distributed as $N(0, \sigma^2)$, which are also independent with U_{0j}.

With this model specification in Eq. (2.2), the covariance between any two individuals i and i' within any cluster j can be calculated as follows:

$$
\begin{aligned}
Cov\left(y_{ij}, y_{i'j}\right) \\
&= Cov\left(\beta_0 + U_{0j} + \beta_1 x_{1ij} + \epsilon_{ij}, \beta_0 + U_{0j} + \beta_1 x_{1i'j} + \epsilon_{i'j}\right) \\
&= Cov\left(U_{0j} + \epsilon_{ij}, U_{0j} + \epsilon_{i'j}\right) \\
&= Cov\left(U_{0j}, U_{0j}\right) + Cov\left(U_{0j}, \epsilon_{i'j}\right) + Cov\left(\epsilon_{ij}, U_{0j}\right) + Cov\left(\epsilon_{ij}, \epsilon_{i'j}\right) \\
&= Cov\left(U_{0j}, U_{0j}\right) \text{ (All three terms are zero due to independence)} \\
&= \tau^2.
\end{aligned}
\tag{2.3}
$$

In addition, with Eq. (2.2), the variance for any individual i or i' can be shown as

$$
\begin{aligned}
Var(y_{ij}) &= Var(\beta_0 + U_{0j} + \beta_1 x_{1ij} + \epsilon_{ij}) \\
&= Var(U_{0j} + \epsilon_{ij}) = \tau^2 + \sigma^2.
\end{aligned}
\tag{2.4}
$$

Therefore, the correlation between any two individuals i and i' within any cluster j can be obtained as follows:

$$
cor(y_{ij}, y_{i'j}) = \frac{Cov(y_{ij}, y_{i'j})}{\sqrt{Var(y_{ij})Var(y_{i'j})}} = \frac{\tau^2}{\tau^2 + \sigma^2} = \rho_I = ICC.
\tag{2.5}
$$

This explains why ICC is the correlation between any two individuals within any cluster.

2.2.3   How to Estimate the ICC

To estimate ρ_I, we need to estimate τ^2 and σ^2 from the sample data, which can be done as follows.

2.2.3.1 Estimate Within-Cluster Variance of σ^2

Like in linear regression, the within-cluster variance σ^2 is estimated as the sum of squares errors, which is defined as the weighted average of within-cluster variances as

$$\hat{\sigma}^2 = \frac{\sum_{j=1}^{J}(n_j - 1)S_j^2}{N - J}, \tag{2.6}$$

where S_j^2 is the sample variance within cluster j and calculated as $S_j^2 = \frac{\sum_{i=1}^{n_j}(y_{ij} - \bar{y}_j)^2}{n_j - 1}$, where n_j is the sample size of cluster j. N is the total sample size as $N = n_1 + n_2 + \cdots + n_J$ with J is the total number of clusters.

2.2.3.2 Estimate the Between-Cluster Variance τ^2

Due to the correlation and random variation within each cluster, the estimation of the between-cluster variance τ^2 should remove this within-cluster variation. Therefore, we first calculate the weighted between-cluster variance and then remove the random variation within clusters as follows:

1. Calculate the weighted between-cluster variance:

$$\hat{S}_B^2 = \frac{\sum_{j=1}^{J} n_j(\bar{y}_j - \bar{y})^2}{\tilde{n}(J - 1)}, \tag{2.7}$$

where $\bar{y}_j$ is the mean on the response variable for cluster j and $\bar{y}$ is the overall mean on the response variable from all clusters and

$$\tilde{n} = \frac{1}{J - 1}\left(N - \frac{\sum_{j=1}^{J} n_j^2}{N}\right). \tag{2.8}$$

2. Remove the within-cluster random variation.

S_B^2 cannot be used as the final estimate of τ^2 due to the random variation among subjects within the same cluster, which should be removed. Therefore, the final between-cluster variance is estimated as follows:

$$\hat{\tau}^2 = S_B^2 - \frac{\hat{\sigma}^2}{\tilde{n}}. \tag{2.9}$$

2.2.3.3 Estimate ρ_I

With the estimated $\hat{\sigma}^2$ and τ^2, the *ICC* can be estimated as

$$\hat{\rho}_I = \frac{\hat{\tau}^2}{\hat{\tau}^2 + \hat{\sigma}^2}. \tag{2.10}$$

2.2.3.4 Importance of *ICC*

The *ICC* is a very important measure for multi-level data structures and is also a very important parameter in multi-level regression to be estimated. As mentioned before, *ICC* measures how much correlation exists in multi-level data and indicates how much multi-level data structure may impact the outcome variable in multi-level modeling.

Large *ICC* values indicate that between-group variations are greater than within-group variations. Hence, the larger the *ICC* values, the greater the impact of clustering on multi-level modeling. As such, a large *ICC* value indicates a need to use multi-level modeling in data analysis. However, there is no consensus as to how large an *ICC* value should be to warrant a MLM, although a double-digit (in percentage) *ICC* generally indicates that a MLM should be used.

2.2.4 Why We Cannot Ignore Multi-Level Data Structure

Simply put, from a statistical modeling point of view, ignoring the multi-level data structure would naively assume that the data are independent, while they are not. This would produce problems and erroneous regression analysis. Specifically as below:

2.2.4.1 Erroneous Statistical Inference

Due to its nested structure, the data within clusters are correlated and, therefore, dependent. If the multi-level nature of the data is ignored and a classical linear regression is used to analyze the correlated data, the standard error of the parameter estimates will be wrong. Consequently, the significance tests associated with the parameters will be incorrect. This can be obvious since the standard errors from the classical linear regression assume independence of observations.

More specifically, the standard errors will be mostly underestimated. Underestimation of these standard errors would lead to a large t-statistic and a smaller p-value for the regression. This would increase the potential for type-I error in which the null hypothesis is wrongfully rejected, resulting in false positive findings in the research.

2.2.4.2 Model Misspecification

Model misspecification can be another consequence of ignoring multi-level data structures. Ignoring multi-level data structures can lead to ignoring important between-level relationships, which could either be linear or nonlinear. For example, in the two-level educational intervention where students (in level 1) are nested within schools (in level 2), if ignoring information at the school level, such as school administration and teachers' education, etc., we may overlook some important school-level variables that influence performance at the student level. This influence can be linearly or mostly nonlinearly related. Therefore, ignoring multi-level data structures would lead to overlooking important nonlinear relationships and model misspecification. Model misspecification would also lead to erroneous research findings.

2.3 Why Random-Intercept and Random-Slope MLM

To understand random-intercept and random-slope MLM intuitively, we first illustrate them using the dataset from *Public Examination Scores from 649 students in 73 schools* in Sect. A.2.

First, we load the data into *R* using the following code chunk:

```
# Read in the data#
dSci = read.csv("Sci.csv", header=T)
# Check with the dimension of the data
dim(dSci)
```

```
## [1] 1905     5
```

```
# Check how many students
length(unique(dSci$ID))
```

```
## [1] 649
```

```
# Check how many centre in this data
length(unique(dSci$Centre))
```

```
## [1] 73
```

```
# Print the first 6 observations
head(dSci)
```

```
##    Centre ID Gender Written Course
## 1   20920 16      0      38     22
## 2   20920 25      1      60     77
## 3   20920 27      1      63     83
## 4   20920 31      1      59     95
## 5   20920 42      0      27     48
## 6   20920 62      1      58     86
```

```
# Data Summary
summary(dSci)
```

```
##         Centre              ID                Gender
##    Min.   :20920    Min.    :    1    Min.     :0.0000
##    1st Qu.:60501    1st Qu.:   64    1st Qu.:0.0000
##    Median :68133    Median :  133    Median :1.0000
##    Mean   :62128    Mean    :1037    Mean     :0.5921
##    3rd Qu.:68411    3rd Qu.:  458    3rd Qu.:1.0000
##    Max.   :84772    Max.    :5521    Max.     :1.0000
##         Written             Course
##    Min.    :   1.00   Min.    :  10.00
##    1st Qu.:  62.00   1st Qu.:  68.00
##    Median :  75.00   Median :  82.00
##    Mean    :  74.94   Mean    :  79.03
##    3rd Qu.:  89.00   3rd Qu.:  92.00
##    Max.    :144.00   Max.    :108.00
```

Second, we select a subset of data from 9 schools to illustrate and investigate the possible linear relationship and the associated intercepts and slopes from each school. We use the following *R* code chunk:

```
# Subset schools
subsch = sort(unique(dSci$Centre))[1:9]
# Get the data for these schools and name it as "dat"
dat = dSci[dSci$Centre %in% subsch,c("Centre","Written","Course")]
dat$Centre = as.factor(dat$Centre)
```

Third, we plot the data as in Fig. 2.1 from these schools to examine the linear relationship between the written paper score (i.e., *Written*) and the score of coursework evaluated by their teachers (i.e., *Course*). We also include a linear regression for each school as follows:

```
# load the lattice package
library(lattice)
# Call xyplot and add regression lines
xyplot(Course~Written|Centre,type=c("p","r"), lwd=3, dat)
```

These plots show linear relationships between the score of the course and the score from their written papers for students among schools, with some of them positively related and some of them negatively related. Among the linear relationships from the 9 schools, the intercepts and slopes are different for each school. This leads to the so-called random-intercept and random-slope multi-level modeling.

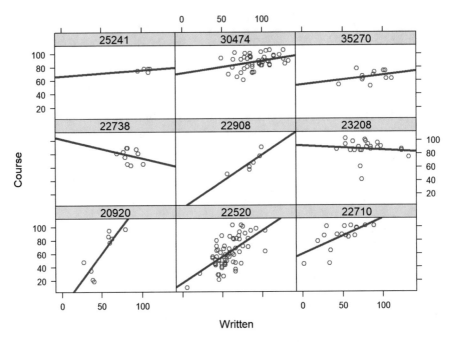

Fig. 2.1 Relationship between Written and Course for each school

To make this clearer, we can plot the data from these schools together as in Fig. 2.2 using the following R code chunk:

```
par(mar = c(4, 4, .1, .1))
xyplot(Course~Written, group=Centre, type=c("p","r"),
    auto.key = list(columns = nlevels(dat$Centre)), dat)
```

This plot shows each school's different intercepts and slopes. If a typical regression were used to analyze this data, the different intercepts and slopes would be ignored. Such a model would poorly estimate relationships among variables.

We can plot the data from all 73 schools together as shown in Fig. 2.3 as follows:

```
par(mar = c(4, 4, .1, .1))
xyplot(Course~Written, group=Centre, type=c("p","r"), dSci)
```

With all the data together, we can see why a random-intercept and random-slope multi-level model (commonly called *mixed-effects model*) is needed in order to incorporate the within-school and between-school variabilities together for comprehensive modeling.

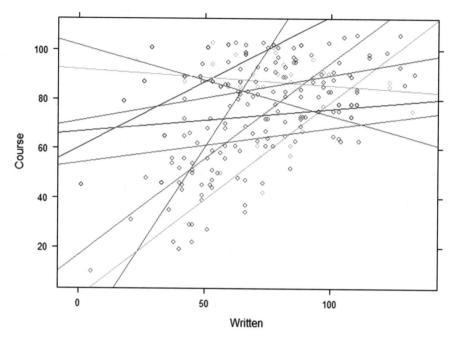

Fig. 2.2 Relationship between Written and Course for nine schools

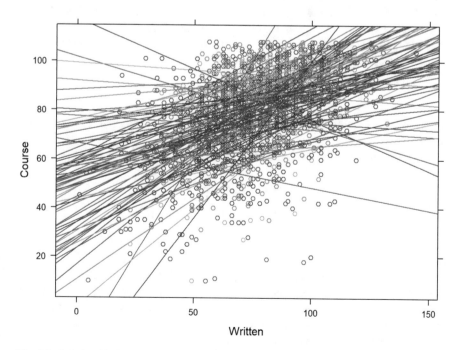

Fig. 2.3 Relationship between Written and Course for all schools

2.4   Two-Level MLM

As illustrated in Fig. 2.3, the real data analysis should have started with both the random-intercept and random-slope models. However, for model building, based on the principle of model parsimony, we typically start from the simplest model and build it up to the most complicated model. We then select the best model based on statistical model selection criteria, such as the likelihood ratio test.

In this section, we will describe this process from the *null model* (sometimes, it is called *mean model*), to the random-intercept only model (in short, *random-intercept model*), and to both random-intercept and random-slope models (in short, *random-slope model*). All these models are termed as mixed-effects models in statistical modeling since they are a mixture of both fixed-effects and random-effects components.

For simplicity, we will begin by describing these models with one independent variable x_1 since this approach can be easily expanded to include multiple independent variables.

2.4.1   NULL Model

Generally, the null model is the first model used when estimating a multi-level model. The model contains no predictors, and its purpose is to determine if differences exist, on average, in the outcome variable (i.e., the score for the *Course* in *dSci* dataset) across levels. For this reason, the null model is also called the mean model. The model is formulated as follows:

$$y_i = \beta_0 + \epsilon_i, \tag{2.11}$$

where y_i is the outcome for the ith individual (i.e., y_i = score of the course for student i).

With a multi-level data structure, the above model would be

$$\text{Level 1}: \quad y_{ij} = \beta_{0j} + \epsilon_{ij} \tag{2.12}$$

$$\text{Level 2}: \quad \beta_{0j} = \beta_0 + U_{0j}, \tag{2.13}$$

where i is the ith individual (i.e., student in the *dSci* dataset) and j for cluster (i.e., class in the *dSci* dataset).

In this null model, β_0 represents the grand mean/average or general intercept value that holds across clusters, which is a fixed effect because it is fixed to a constant across all clusters. U_{0j} is a cluster-specific effect on the intercept. Therefore, it is a random effect because it varies from cluster to cluster with a mean of 0 (as the deviation from the fixed effect) and a between-cluster variance τ^2. The error term ϵ_{ij} denotes the within-cluster variation with within-cluster variance of σ^2. In MLM, we further assume U_{0j} and ϵ_{ij} are uncorrelated.

Put together, the null model would be

$$y_{ij} = \beta_0 + U_{0j} + \epsilon_{ij}, \tag{2.14}$$

where $U_{0j} \sim N(0, \tau^2)$, $\epsilon_{ij} \sim N(0, \sigma^2)$, and U_{0j} and ϵ_{ij} are independent.

2.4.2   Random-Intercept Only Model

The random-intercept only model (in short, random-intercept model) extends the null model by including predictors. The random-intercept model assumes each cluster is unique. Therefore, using our example, the course scores *Course* (denoted by y_{ij}) of each individual student are predicted by the varied group intercepts. Said another way, the uniqueness of each cluster will impact the cluster's student scores on written paper *Written* (denoted by x_{1ij}). The model is as follows:

$$\text{Level 1}: \ y_{ij} = \beta_{0j} + \beta_{1j}x_{1ij} + \epsilon_{ij}$$
$$\text{Level 2}: \ \beta_{0j} = \beta_0 + U_{0j}$$
$$\beta_{1j} = \beta_1.$$

The rest of the assumptions are the same in the null model. The general model is then

$$y_{ij} = \beta_{0j} + \beta_{1j}x_{1ij} + \epsilon_{ij} = \beta_0 + U_{0j} + \beta_1 x_{1ij} + \epsilon_{ij} \tag{2.15}$$

with same assumption on error terms that $U_{0j} \sim N(0, \tau^2)$ and $\epsilon_{ij} \sim N(0, \sigma^2)$.

Keep in mind that in the random-intercept MLM model, the intercepts will be random (or change as seen in Fig. 2.3), but the slopes are assumed to be the same for all the clusters. In reality, the slopes are different as seen in Fig. 2.3, which leads to random-intercept and random-slope MLMs.

2.4.3   Random-Slope Model

Again, in the above random-intercept model, we assume the impact of the independent variable on the outcome variable, as measured by β_1, is constant to all clusters. This assumption may not be true in some situations as seen in Fig. 2.3. To incorporate the different impacts between clusters, we assume that the cluster-specific slope β_{1j} changes from cluster to cluster as a random effect $\beta_1 + U_{1j}$ with a mean of β_1. The random component U_{1j} varies from cluster to cluster with a mean

0 and a variance τ_1^2. Furthermore, β_{10} is the average relationship of x with y across clusters, and U_{1j} is the cluster-specific variation of the relationship between these two variables.

The random-slope model extends the random-intercept model by including random effects for slopes as follows:

$$\text{Level 1}: \quad y_{ij} = \beta_{0j} + \beta_{1j}x_{ij} + \epsilon_{ij}$$

$$\text{Level 2}: \quad \beta_{0j} = \beta_0 + U_{0j}$$

$$\beta_{1j} = \beta_1 + U_{1j}.$$

The general model is then

$$y_{ij} = \beta_{0j} + \beta_{1j}x_{ij} + \epsilon_{ij}$$

$$= \beta_0 + U_{0j} + \beta_1 x_{ij} + U_{1j}x_{ij} + \epsilon_{ij}. \tag{2.16}$$

In this model, $\beta_0 + \beta_1 x_{ij}$ is the fixed-effects component and $U_{0j} + U_{1j}x_{ij} + \epsilon_{ij}$ is the random-effects component, and $U_{1j}x_{ij}$ indicates an interaction between cluster and x, such that the relationship of x and y is not constant across clusters.

Similar to the random-intercept model, the random-slope model can model the between-cluster intercept variance as denoted by τ_0^2. In contrast to the random-intercept model, the random-slope model also incorporates the between-cluster slopes (impacts) variance. To differentiate these two sources of between-group variance, we now denote the variance of U_{0j} as τ_0^2 and the variance of U_{1j} as τ_1^2.

We expect these two random components to be correlated and to have a correlation ρ_{01} and a covariance of $\tau_{01} = \tau_0\tau_1\rho_{01}$. However, we assume these random components are independent with the error ϵ_{ij}.

Putting all these together, the random-slope model is formulated as follows:

$$\text{Level 1}: \quad y_{ij} = \beta_{0j} + \beta_{1j}x_{ij} + \epsilon_{ij}$$

$$\text{Level 2}: \quad \beta_{0j} = \beta_0 + U_{0j}$$

$$\beta_{1j} = \beta_1 + U_{1j}$$

with error distributions:

$$\text{Within-Cluster Errors}: \quad \epsilon_{ij} \sim N(0, \sigma^2)$$

$$\text{Between-Cluster Errors}: \quad \begin{pmatrix} U_{0j} \\ U_{1j} \end{pmatrix} \sim \left(\begin{pmatrix} 0 \\ 0 \end{pmatrix}, \begin{pmatrix} \tau_0^2 & \tau_0\tau_1\rho_{01} \\ \tau_0\tau_1\rho_{01} & \tau_1^2 \end{pmatrix} \right).$$

2.5 Three-Level MLMs

With the understanding of two-level MLM, the three-level MLM can be similarly discussed by incorporating an additional higher level. The three-level MLM is defined as follows:

$$\text{Level 1: } y_{ijk} = \beta_{0jk} + \beta_{1jk}x_{ijk} + \epsilon_{ijk}, \tag{2.17}$$

where the subscript k represents the level 3 cluster to which the second level belongs. In this formulation, the third level is $k = 1, \cdots, K$, where K is the total number of clusters in the third level. In the second level, the cluster $j = 1, \cdots, J_k$ with J_k as the total number of clusters in the second level for the jth cluster (i.e., the total number of clusters in level 2 is $J = \sum_{k=1}^{K} J_k$). Finally, in the first level, $i = 1, \cdots, n_{jk}$ with the total number of observations in the jth level 2 cluster that is nested within the kth level 3 cluster. Therefore, the total number of observations $N = \sum_{k=1}^{K} \left[\sum_{j=1}^{J_k} n_{jk} \right]$.

For example, in a three-level educational intervention, students (i.e., level 1, denoted by i) nested within classrooms (i.e., level 2, denoted by j) nested within school (i.e., level 3, denoted by k). Suppose that there are 5 schools in the level 3 (i.e., $k = 1, \cdots, K = 5$) with the number of classrooms of $(3, 4, 2, 5, 2)$ at level 2 for each school. Then, there are $J_1 = 3$ classrooms for the first school, $J_2 = 4$ classrooms for the second school, $J_3 = 2$ classrooms for the third school, $J_4 = 5$ classrooms for the fourth school, and $J_5 = 2$ classrooms for the fifth school with total number of classrooms $J = J_1 + J_2 + J_3 + J_4 + J_5 = 3 + 4 + 2 + 5 + 2 = 16$. To simplify our calculation, we assume that there are equal number of students (say, $n_{jk} = 20$ students in each classroom) in the level 1 nested within each of these 16 classrooms. Therefore, the total number of observations $N = \sum_{k=1}^{K} \left[\sum_{j=1}^{J_k} n_{jk} \right] = 16 \times 20 = 320$.

To formulate three-level MLMs, we would have to specify different levels of random-intercepts and random-slopes, i.e., whether the slopes and intercepts are random at both levels 2 and 3 or only at one level. Of course, the specification of slopes and intercepts are always based on intervention theory and the research question to be answered.

Typically, MLM models assume that the level 1 intercepts and slopes are random at both levels 2 and 3 as described as follows. This specification allows for the most complex model possible for a three-level data structure.

The general specification for random-slope and random-intercept in both level 2 and level 3 is as follows:

$$\text{Level 2}: \beta_{0jk} = \beta_{00k} + U_{0jk}$$
$$\beta_{1jk} = \beta_{10k} + U_{1jk}$$
$$\text{Level 3}: \beta_{00k} = \beta_0 + V_{00k}$$
$$\beta_{10k} = \beta_1 + V_{10k}.$$

Simple substitution is used to obtain the expression for the level 1 intercept and slope in terms of both level 2 and level 3 parameters

$$\beta_{0jk} = \beta_0 + V_{00k} + U_{0jk}$$
$$\beta_{1jk} = \beta_1 + V_{10k} + U_{1jk}.$$

In turn, these terms may be substituted into the equation to provide the full three-level MLM

$$y_{ijk} = \beta_0 + V_{00k} + U_{0jk} + (\beta_1 + V_{10k} + U_{1jk})x_{ijk} + \epsilon_{ijk}. \qquad (2.18)$$

The specification of random effects of the three-level MLM is more complicated than that of the two-level MLM. We will come back for a more detailed discussion in Chap. 4.

2.6 Longitudinal Data and Their Relationship to MLMs

2.6.1 Longitudinal Data Can be Modeled as MLMs

In longitudinal studies, data are collected from the same individual at multiple points in time. For example, in an educational intervention study, we might measure the scores of students every semester for several years.

With this longitudinal design, we would be able to investigate the growth and change in score over time. Such models are sometimes called growth models, and they align with a MLM framework. For example, in a typical two-level longitudinal MLM model, students would represent the level 2 (cluster) variable, and the longitudinal measurement of individual test scores would represent level 1. Similarly, if students are nested within schools, we would have a three-level MLM model, with school serving as the third level.

2.6.2 Special Error Structure for Longitudinal Data

Unlike other MLM applications, the error terms in longitudinal MLM data can take specific forms. These error terms reflect the way in which measurements made over time relate to one another, and they are typically more complex than the basic error structure described thus far. We will discuss the details of longitudinal data analysis in Chap. 5.

2.7 MLM Assumptions and Parameter Estimation Methods

2.7.1 Assumptions Underlying MLMs

Similar to classical linear regression, several assumptions must be met to appropriately use MLMs. If these assumptions are not met, the model parameter estimates may be untrustworthy.

Generally, MLM and single-level multiple linear regression (MLR) model have some different assumptions due to multi-level data structure. For example, in two-level MLM, we typically assume:

1. The level 2 residuals are independent between clusters. This requires that the random-intercept and random-slope(s) at level 2 are independent from each other across clusters.
2. The level 2 intercepts and slopes are independent of the level 1 residuals. This means that the errors for the cluster-level estimates are not related to the errors at the individual level.
3. The level 1 residuals are normally distributed and have constant variances. This assumption is similar to the assumption in the classical linear regression model.
4. The level 2 intercept and slope(s) have a multivariate normal distribution with a constant covariance matrix.

Each of these assumptions can be directly assessed from the data and model residuals. The methods for checking the MLM assumptions are very similar to those of linear regression discussed in Chap. 1.

2.7.2 Methods for Parameter Estimation

We discussed the ordinary least squares estimation (i.e., *LSE*) in Chap. 1 for parameter estimation in classical linear regression. Due to the multi-level data structure, the ordinary *LSE* cannot be used anymore to estimate parameters from MLMs. Instead, the maximum likelihood estimation (*MLE*) and restricted maximum likelihood (*REML*) should be used as discussed in Pinheiro and Bates (2000).

2.7.2.1 Maximum Likelihood Estimation

In *MLE*, we estimate the parameters by maximizing the likelihood of obtaining the sample that we obtained. With the multi-level data structure, we formulate a likelihood function based on the MLM model and its distribution assumptions at different levels with the observed data. This likelihood function is then maximized with numerical optimization.

Different from the ordinary *LSE* where we can get an analytical formula, i.e., in Eq. (1.7) to calculate the parameter estimates $\hat{\beta}$, the maximum likelihood estimation is unique as follows:

- *MLE* has no closed-form solution generally. Therefore, an iterative numerical optimization algorithm must be used to search for parameter values that maximize the likelihood function.
- *MLE* is computationally intensive, especially for complex models and large samples.
- *MLE* is not guaranteed to obtain the parameter estimates and the search algorithm can be diverged without solution.
- *MLE* is not guaranteed to obtain the parameter estimates you desired due to local optimization of most of the search algorithms we implemented. In this situation, different initial values are recommended for the numerical search algorithm.

2.7.2.2 Restricted Maximum Likelihood Estimation

A variant of *MLE*, the restricted maximum likelihood estimation (*REML*), is more accurate in estimating variance parameters due to its different adjustments in calculating degrees of freedom. *REML* is typically the default estimation method in MLM implemented in *R*.

To determine the appropriate degrees of freedom, *REML* estimates variance components by accounting for the number of parameters being estimated. Therefore, the *REML* takes into account the loss in degrees of freedom due to estimating fixed effects. In contrast, *MLE* does not account for this loss, and therefore, the variance estimations in *MLE* are biased and underestimated. Although *MLE* estimates are biased, this bias could be small with a larger number of level 2 clusters. So, the difference between *MLE* and *REML* becomes very small and negligible as the number of level 2 clusters increases.

REML is generally the preferred method for estimating multi-level models, which is why *REML* is implemented as the default estimation method to obtain variance estimates. Although *REML* is suggested over *MLE* in estimating variance components, *MLE* should be used for statistical testing, especially for variance parameters in any random effect. More discussion can be found in Pinheiro and Bates (2000).

2.8 Exercises

Use the description in Sect. 2.3 with the dataset from *Public Examination Scores from 649 students in 73 schools* in Sect. A.2:

1. formulate a two-level MLM to model the score of the course (i.e., *Course*) and their written score (i.e., *Written*) with students (i.e., *ID*) as level 1 and school as level 2 (i.e., *Centre*),
2. make plots to show the difference between *Gender*.

Chapter 3
Two-Level Multi-Level Modeling

With the introduction and conceptualization of multi-level modeling (MLM) in Chap. 2, we will focus on the detailed model building and implementation in R in this chapter.

First, we will describe model building for two-level MLM from the null model, to a random-intercept MLM, and then to a random-slope MLM. We again use the dataset from Public Examination Scores from 649 students in 73 schools in Appendix A.2 to illustrate model fitting with the R package *nlme* (Pinheiro and Bates 2000).

Note to readers: We will use R package *nlme* (i.e., *non*linear *m*ixed-*e*ffects model) in this chapter. Remember to install this R package to your computer using *install.packages("nlme")* and load this package into R session using *library(nlme)* before running all the R programs in this chapter.

3.1 The Two-Level MLM

As described in Chap. 2, the general two-level MLM (i.e., linear mixed-effects model) with both random-intercept and random-slope is as follows:

$$Level\ 1: \quad y_{ij} = \beta_{0j} + \beta_{1j}x_{ij} + \epsilon_{ij}$$

$$Level\ 2: \quad \beta_{0j} = \beta_0 + U_{0j} \text{ (Random-intercept)}$$

$$\beta_{1j} = \beta_1 + U_{1j} \text{ (Random-slope)}, \quad (3.1)$$

where j denotes clusters from $j = 1, \cdots, J$ (i.e., J is the total number of clusters), and i $(i = 1, \cdots, n_j$, and n_j is the number of observations in the jth cluster)

© The Author(s), under exclusive license to Springer Nature Switzerland AG 2021
D.-G. (Din) Chen, J. K. Chen, *Statistical Regression Modeling with R*,
Emerging Topics in Statistics and Biostatistics,
https://doi.org/10.1007/978-3-030-67583-7_3

denotes the ith observation from the jth cluster. Therefore, the total number of observations is $N = \sum_{j=1}^{J} n_j$.

In this formulation, the error distributions are assumed as follows:

$$\text{Within-Cluster Error: } \epsilon_{ij} \sim N(0, \sigma^2)$$

$$\text{Between-Cluster Error: } \begin{pmatrix} U_{0j} \\ U_{1j} \end{pmatrix} \sim N\left(\begin{pmatrix} 0 \\ 0 \end{pmatrix}, \begin{pmatrix} \tau_0^2 & \tau_0\tau_1\rho_{01} \\ \tau_0\tau_1\rho_{01} & \tau_1^2 \end{pmatrix} \right).$$

In this model, $\beta_0 + \beta_1 x_{ij}$ is the fixed-effects component and $U_{0j} + U_{1j}x_{ij} + \epsilon_{ij}$ is the random-effects component. Additionally, $U_{1j}x_{ij}$ indicates an interaction between cluster and x, which indicates that the relationship of x and y is not constant across clusters.

So in two-level MLM, we estimate the overall fixed-effects parameters for intercept β_0 and slope β_1, along with the within-cluster standard deviation σ, the between-cluster standard deviations for the random-intercept τ_0 and random-slope τ_1 as well as the correlation between random-intercept and random-slope ρ_{01}.

3.2 MLM Using *R* Package *nlme*

3.2.1 *Data on* Public Examination Scores in 73 Schools

We continue to use the dataset from *Public Examination Scores from 649 students in 73 schools* in Appendix A.2. Notice that the schools are called *Centre* in this dataset, and we will use *School* and *Centre* interchangeably in the analysis of this data in this chapter.

Again let us load the data into *R* using the following code chunk:

```
# Read in the data#
dSci = read.csv("Sci.csv", header=T)
```

We can check the dimensions of the data to see how many observations and how many variables in the dataset as follows:

```
# Check data dimension
dim(dSci)
```

```
## [1] 1905    5
```

We can also count how many students and how many schools (i.e., *Centre*) are in this dataset using the *R* function *length* to count how many *unique* students (i.e., *ID*) and schools (i.e., *Centre*) are in the dataset as follows:

```
# Check how many students
length(unique(dSci$ID))
```

```
## [1] 649
```

```
# Check how many centres in this data
length(unique(dSci$Centre))
```

```
## [1] 73
```

As seen from this output, we know that there are 649 students nested in 73 schools in this dataset.

To further display the nested structure of students (with *Gender* with 0 = boy and 1 = girl) nested within classes in the data, we can call *R* function *xtabs* with format function *ftable* as follows:

```
# Check the nested data structure and print the data for 10 schools
head(ftable(xtabs(~Centre+Gender, dSci)),10)
```

```
##
##              "Gender" "0" "1"
##   "Centre"
##   "20920"           4    5
##   "22520"          22   43
##   "22710"           7   11
##   "22738"           6    4
##   "22908"           1    5
##   "23208"           6   17
##   "25241"           4    1
##   "30474"          23   22
##   "35270"           7    5
##   "37224"          25   19
```

We used *R* function *head* to output the first 10 *Centres* from the total 73 *Centres* to reduce the output. As seen from the output, there are schools with only a few students, but most of the schools have students in the double digits. The data summary can be also seen as follows:

```
# Data Summary
summary(dSci)
```

```
##       Centre            ID              Gender
##   Min.   :20920   Min.   :   1   Min.   :0.0000
##   1st Qu.:60501   1st Qu.:  64   1st Qu.:0.0000
##   Median :68133   Median : 133   Median :1.0000
##   Mean   :62128   Mean   :1037   Mean   :0.5921
##   3rd Qu.:68411   3rd Qu.: 458   3rd Qu.:1.0000
##   Max.   :84772   Max.   :5521   Max.   :1.0000
##      Written          Course
##   Min.   : 1.00   Min.   : 10.00
```

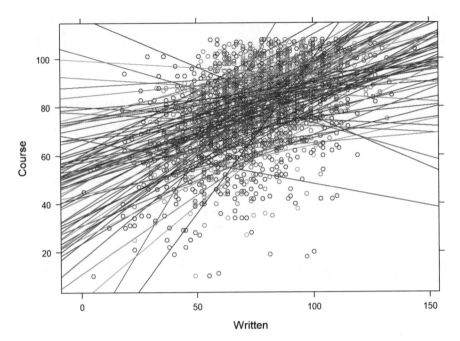

Fig. 3.1 Relationship between Written and Course for each centre

```
##   1st Qu.:  62.00    1st Qu.:  68.00
##   Median :  75.00    Median :  82.00
##   Mean   :  74.94    Mean   :  79.03
##   3rd Qu.:  89.00    3rd Qu.:  92.00
##   Max.   : 144.00    Max.   : 108.00
```

3.2.2 Graphical Analysis

We would like to determine the extent to which the scores on written paper (i.e., *Written*) predict the overall score in the coursework (i.e., *Course*), as evaluated by their teachers.

We first investigate whether a linear relationship between *Written* and *Course* is appropriate for students among schools, and we make use of the *lattice* package to plot the data as follows:

```
# Load the lattice package
library(lattice)
# Sort the data for better plotting
dSci= dSci[order(dSci$Centre, dSci$Written),]
# Make the plot
xyplot(Course~Written, group=Centre, type=c("p","r"), dSci)
```

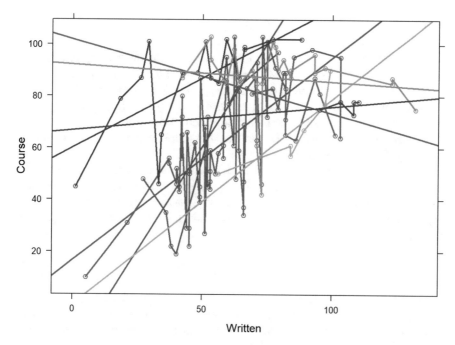

Fig. 3.2 Relationship between Written and Course for 7 Centres

Figure 3.1 plots all 73 centres, and it is difficult to see the real relationship. To make it clearer, let us show the relationship for a few centres:

```
# Make the plot
xyplot(Course~Written, group=Centre, type=c("p","l","r"),
       lwd=2, dSci[dSci$Centre< 30000,])
```

In Fig. 3.2, we used the plotting option *type=c("p", "l", "r")* to plot the data points linked with lines and overlaid with the regression lines. We also used *lwd=2* to double the line width. As seen from this figure, a linear regression for each school is generally reasonable. In addition, this plot showed that each school has a different intercept and slope where most of the schools have positive relationships, but some have negative relationships.

If a typical linear regression were to be used to analyze this data for all schools, the different intercepts and slopes would be ignored. Such a model would poorly estimate the true relationships among variables. Since students are nested within centres (i.e., schools), standard linear regression models are not appropriate anymore. We need MLM to analyze this nested data. As such, we will fit a series of MLMs to determine which model best fits the data.

3.2.3 Response Feature Analysis

We start with the *response feature analysis* to extract the related response features
from the data so we can reasonably formulate a MLM model for this data.

Specifically, the main purpose of this *response feature analysis* for this data
is to extract fundamental features from each *centre* (i.e., school) for simple and
preliminary data exploration and summarization. This feature analysis provides
basic summary information from the data for simple conclusions in addition to
providing directions for further analysis.

From Fig. 3.1, a linear regression model for each school seems reasonable. We
have a sense that we need a model to incorporate the different intercepts and slopes
for each school. That is to say, for each school j, the score for student (i) of *Written*
paper and their score of the *Course* can be reasonably modeled as linear regression
model:

$$y_{ij} = \beta_{0j} + \beta_{1j}x_{ij} + \epsilon_{ij}, \tag{3.2}$$

where y_{ij} denotes the score of *Course* for student i from jth school, β_{0j} and β_{1j}
represent the parameters of intercept and slope in the linear model for the jth school,
and ϵ_{ij} is the error term that is assumed to be normally distributed with mean 0 and
within-school standard deviation of σ.

To extract the response features, we estimate the *slope* (i.e., β_{1j}) and *intercept*
(i.e., β_{0j}) for each school as depicted in Fig. 3.1. To do so, we can loop over the
73 schools to fit a linear regression and then extract the intercepts and slopes (i.e.,
features) to make a dataframe as follows:

```
# Extract the IDs for all School centre
School.IDs = unique(dSci$Centre)
# Number of School
num.Sch = length(School.IDs)
# initiate the intercept and slope
beta0 = beta1 = numeric(num.Sch)

# loop-over to track the beta0 and beta1 for each school
for(j in 1:num.Sch){
# fit regression model for all Schools
mod = lm(Course~Written, dSci[dSci$Centre==School.IDs[j],])
# extract the intercept and slope
beta0[j] = coef(mod)[1];
beta1[j] = coef(mod)[2]
} # end of the j-loop

# make a dataframe "dat.coef"
dat.coef=data.frame(sID=School.IDs, beta0=beta0, beta1=beta1)
# Print the data for the first 6 to see the beta0 and beta1
head(dat.coef)
```

```
##      sID       beta0           beta1
## 1 20920  -19.1470175   1.59635469
## 2 22520   16.8517962   0.78193921
## 3 22710   60.2125581   0.53634478
## 4 22738  101.8383426  -0.28447743
## 5 22908   -0.2958432   0.79327124
## 6 23208   92.1446061  -0.06560847
```

```
# print the summary
summary(dat.coef)
```

```
##        sID                 beta0                    beta1
## Min.    :20920    Min.    :-28.18      Min.    :-0.4150
## 1st Qu.:60455     1st Qu.: 36.99       1st Qu.: 0.1852
## Median :68107     Median : 53.77       Median : 0.3538
## Mean    :61281    Mean    : 52.12      Mean    : 0.3594
## 3rd Qu.:68405     3rd Qu.: 66.64       3rd Qu.: 0.5224
## Max.    :84772    Max.    :117.49      Max.    : 1.5964
```

We can calculate some statistics that are the response features of the MLM in Eq. (3.1) as listed below:

```
# Mean for intercepts, see Fig. 3.1
mean(dat.coef[,2])
```

```
## [1] 52.11768
```

```
# Mean for the slopes, see Fig. 3.1
mean(dat.coef[,3], na.rm=T);
```

```
## [1] 0.3594173
```

```
# Covariance matrix for intercepts and slopes
va = var(dat.coef[,2:3], na.rm = T);va
```

```
##              beta0         beta1
## beta0 628.304396  -6.99316440
## beta1  -6.993164   0.09429743
```

```
# Standard deviations
sqrt(va)
```

```
## Warning in sqrt(va): NaNs produced
```

```
##           beta0      beta1
## beta0 25.066        NaN
## beta1    NaN 0.3070789
```

```
# Correlations
cov2cor(va)
```

```
##                  beta0           beta1
## beta0    1.0000000    -0.9085289
## beta1   -0.9085289     1.0000000
```

Notice that from the above summary outputs for the dataframe *dat.coef*, the mean intercept of 52.118 should be an estimate and close to β_0 and the mean slope of 0.359 should be an estimate and close to the β_1 in the MLM (3.1). In addition, the calculated variance–covariance matrix $\begin{pmatrix} 628.304 & -6.993 \\ -6.993 & 0.094 \end{pmatrix}$ should be a reasonable estimate of the between-school variance–covariance matrix $\begin{pmatrix} \tau_0^2 & \tau_0\tau_1\rho_{01} \\ \tau_0\tau_1\rho_{01} & \tau_1^2 \end{pmatrix}$, where $\hat{\tau}_0$ is estimated at 25.066, $\hat{\tau}_1$ at 0.307, and $\hat{\rho}_{01}$ at -0.909, if a MLM (3.1) is used. This is the essence of the *response feature analysis*.

A bivariate plot of the *beta0* and *beta1* from these schools appears in Fig. 3.3, which is generated with the following *R* code:

```
# Make histogram for both intercept and slope
beta0.hist = hist(beta0,plot=F)
beta1.hist = hist(beta1,plot=F)

# make layout for plotting
top = max(c(beta0.hist$counts, beta1.hist$counts))
nf  = layout(matrix(c(2,0,1,3),2,2,byrow=T), c(3,1), c(1,3),T)
par(mar=c(5,4,1,1))

# plot the intercept and slope
plot(beta1~beta0,las=1,dat.coef,xlab="Intercept (i.e., beta0)",
   ylab="Slope (i.e., beta1)")

par(mar=c(0,4,1,1))
# add the intercept histogram
barplot(beta0.hist$counts, axes=FALSE,
     ylim=c(0, top), space=0)
par(mar=c(5,0,1,1))

# add the slope histogram
barplot(beta1.hist$counts, axes=FALSE,
     xlim=c(0, top), space=0, horiz=TRUE)
```

Figure 3.3 clearly shows that the intercepts from these 73 schools are about normally distributed with a mean of 52.118 and a standard deviation of 0.307 (the histogram at the top) and the slopes are normally distributed with a mean of 0.359 and a standard deviation of 0.307 (the histogram at the right side). In addition, all the intercepts and slopes are negatively correlated with a correlation coefficient of −0.909. The variations embedded in the intercepts and slopes illustrate the modeling of random effects for both intercepts and slopes.

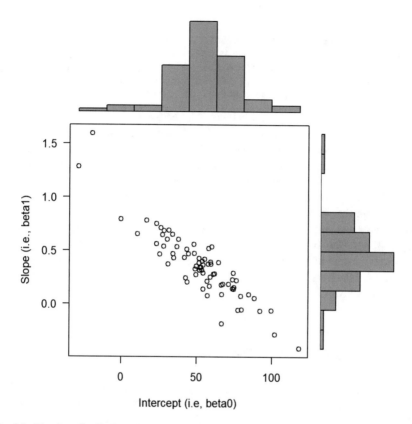

Fig. 3.3 Bivariate distribution of intercepts and slopes

We can further model the slope and intercept relationship with a linear regression as follows:

```
# fit linear regression
mod.coef = lm(beta1~beta0, dat.coef)
summary(mod.coef)
```

```
##
## Call:
## lm(formula = beta1 ~ beta0, data = dat.coef)
##
## Residuals:
##      Min        1Q    Median        3Q       Max
## -0.38092  -0.08674   0.02024   0.07572   0.44375
##
## Coefficients:
##              Estimate Std. Error t value Pr(>|t|)
## (Intercept)  0.9394984  0.0350862    26.78   <2e-16 ***
## beta0       -0.0111302  0.0006075   -18.32   <2e-16 ***
```

```
## ---
## Signif. codes:   0 '***' 0.001 '**' 0.01 '*' 0.05 '.' 0.1 ' ' 1
##
## Residual standard error: 0.1292 on 71 degrees of freedom
## Multiple R-squared:   0.8254, Adjusted R-squared:   0.823
## F-statistic: 335.7 on 1 and 71 DF,   p-value: < 2.2e-16
```

In this model *mod.coef*, the relationship between the *beta1*s and *beta0*s is significant with a negative estimate of -0.011, indicating that there is a significant correlation between the *beta1* and *beta0*, which confirms the graphical illustration in Fig. 3.3. The R^2 in the above model is 0.8254, which is the square of the correlation $R = -0.9085$ and that is the $\hat{\rho}_{01}$ in the MLM (3.1).

In summary, the response feature analysis extracts fundamental features from each stratum (i.e., *centre* in this data) on their intercepts and slopes. These features in intercepts and slopes help us to see the need for a random-intercept and random-slope MLM. However, the analysis loses information since other features are excluded. A more efficient analysis is to use all the information from all the data in a comprehensive manner to capture the multi-level data structure, which is the essence of MLM.

3.2.4 Fitting the Null Model (i.e., the Mean Model)

The mean model is usually the first model used to determine goodness of fit. This model includes no independent variables, and the cluster (e.g., *Centre*) is a random effect.

In this case, the general two-level MLM in Eq. (3.1) is then simplified as the two-level null model equation as follows:

$$y_{ij} = \beta_0 + U_{0j} + \epsilon_{ij}. \tag{3.3}$$

In this null model, we estimate the fixed-effects overall intercept parameter, β_0, and the two random effects of within-cluster variance of σ^2 and between-cluster variance τ_0^2 in the random effects $U_{0j} + \epsilon_{ij}$.

In our example, the model will allow us to see how much variability exists across schools when the mean score of the *Written* is at 0 (hypothetically). When such clustering by *Centre* is considered, the mean model will allow us to obtain estimates of both the residual and random-intercept variance. In addition, the mean model allows us to estimate the *ICC*. The greater the ICC score, the greater the impact of clustering on the outcome variable.

First, we load the library:

```
library(nlme)
```

Detailed explanations of this package *nlme* can be seen in Appendix B, and the associated explanations on the R function *lme* in Appendix B.1.

Next, we fit the null model using the following *R* code chunk:

```
Mod0 = lme(fixed = Course~1, random = ~1|Centre, data =dSci)
# summary of the MLM fitting
summary(Mod0)
```

```
## Linear mixed-effects model fit by REML
##   Data: dSci
##          AIC      BIC    logLik
##     15899.7 15916.36 -7946.852
##
## Random effects:
##   Formula: ~1 | Centre
##           (Intercept) Residual
## StdDev:     9.512016 15.05455
##
## Fixed effects: Course ~ 1
##                  Value Std.Error   DF  t-value p-value
## (Intercept) 79.35806  1.205848 1832 65.81101       0
##
## Standardized Within-Group Residuals:
##        Min         Q1        Med        Q3        Max
## -4.9462905 -0.5186608  0.1137861  0.6810713  2.6983559
##
## Number of Observations: 1905
## Number of Groups: 73
```

As seen in *Mod0*, this null model does not contain independent variables. In the null model *Mod0*, the estimated within-school standard deviation is $\hat{\sigma} = 15.055$ and the between-school standard deviation $\hat{\tau} = 9.512$. The estimated overall mean score of *Course* is $\hat{\beta}_0 = 79.358$.

Nonetheless, we can still use this model to investigate some important information about the structure of the data.

- *Model selection measures*:

The AIC, BIC, and logLik values allow us to compare the null model to other MLMs that include independent variables, which will be discussed later.

- *Estimation of ICC*:

The null model estimates both the within-cluster variance σ^2 and the between-cluster variance τ^2. With these estimates, we can estimate the ICC ρ_I as follows:

```
# Get the within-/between-cluster Covariance matrix from Mod0
estVar0  = VarCorr(Mod0);estVar0
```

```
## Centre = pdLogChol(1)
##              Variance   StdDev
## (Intercept)  90.47845   9.512016
## Residual    226.63957  15.054553
```

```
# Get the between-cluster variance
tausq0    = as.numeric(estVar0[1,1]);tausq0
```

```
## [1] 90.47845
```

```
# Get the within-cluster variance
sigmasq0 = as.numeric(estVar0[2,1]);sigmasq0
```

```
## [1] 226.6396
```

```
# Then estimate ICC
estICC0 = tausq0/(tausq0+sigmasq0)
# print the value
estICC0
```

```
## [1] 0.2853148
```

This value can be interpreted as the overall impact of *Centre* (i.e., schools) clustering on the score of *Course* among students within the same school. It can be seen in this example, the ICC is estimated at 28.53%, which is quite large and indicates significant within-cluster dependence. In this case, the classical linear regression might give biased estimates.

3.2.5 Fitting MLM with Random-Intercept Model

Extending the null model in Eq. (3.3), we can add independent variables to investigate the predictability of the independent variables to the dependent variable. The null model would become

$$y_{ij} = \beta_0 + U_{0j} + \beta_1 x_{ij} + \epsilon_{ij}. \tag{3.4}$$

In this model, an extra fixed-effects term $\beta_1 x_{ij}$ is added to estimate the relationship between the independent variable x (i.e., *Written*) and the outcome variable y (i.e., *Course*). The random-effects term is still the same like in the null model as $U_{0j} + \epsilon_{ij}$.

To include the *Written* score as the independent variable, we can extend the null MLM to a random-intercept MLM as follows:

```
# random-intercept MLM
Mod1 = lme(fixed = Course~Written, random = ~1|Centre, data = dSci)
# print the summary
summary(Mod1)
```

```
## Linear mixed-effects model fit by REML
##  Data: dSci
##         AIC       BIC     logLik
```

```
##   15476.84 15499.04 -7734.418
##
## Random effects:
##  Formula: ~1 | Centre
##          (Intercept) Residual
## StdDev:     8.856977 13.42777
##
## Fixed effects: Course ~ Written
##                 Value Std.Error   DF t-value p-value
## (Intercept) 50.29689 1.7295390 1831 29.0811       0
## Written      0.37786 0.0171783 1831 21.9967       0
##  Correlation:
##          (Intr)
## Written -0.764
##
## Standardized Within-Group Residuals:
##        Min         Q1        Med        Q3        Max
## -5.2512569 -0.5365059  0.1030011  0.6761454  2.9180275
##
## Number of Observations: 1905
## Number of Groups: 73
```

In *Mod1*, the *lme* function calls two parts: the *fixed* effects to estimate the overall effects, and the *random* effects to estimate the between-cluster variability and the default within-cluster variability.

The *fixed* part of *lme* defines the fixed effects as seen in the statement:

$$fixed = Course \sim Written,$$

which defines that the score of the *Course* is predicted with the score of *Written* paper as a fixed effect.

The *random* part of *lme* defines the random effects (i.e., random variability between clusters) in the multi-level structure. In the random-intercept model, the syntax for the intercept is 1 as *random = ~1|Centre* with intercepts randomly varying between *Centre*. This corresponds to the data structure of students nested within *Centre* (i.e., schools).

To determine goodness of fit, we can examine the summary of the random-intercept model in *Mod1*.

From the summary, we can see the following.

- *Better Model Fitting*:

Compared to the null model, the random-intercept model's AIC, BIC, and -logLik statistics dropped from 15899.7, 15916.36, and 7946.852 in the null model (*Mod0*) to 15476.84, 15499.04, and 7734.418 in the random-intercept model (*Mod1*), respectively.

In addition, the estimated within-school standard deviation $\hat{\sigma}$ dropped from 15.055 in *Mod0* to 13.427 in *Mod1*, and the between-school standard deviation $\hat{\tau}$ dropped from 9.512 in *Mod0* to 8.857 in *Mod1* due to the inclusion of *Written* in this model.

We would conclude that the random-intercept model *Mod1* better fits the data than the null model *Mod0*.

- *Parameter Estimates*:

From the *fixed-effects* component, we can see that *Written* is a significant predictor of *Course* ($t = 21.997$, $p < 0.0001$). As *Written* score increases by 1 point, the score of the *Course* increases by 0.378 points.

- R^2:

Similarly, we can estimate the R^2 as the proportion of variance in the outcome variable accounted for at each level of the model. In the context of multi-level modeling, R^2 values can be estimated for each level of the model (Snijders and Bosker 2012).

For level 1, we can calculate

$$R_1^2 = 1 - \frac{\sigma_{M_1}^2 + \tau_{M_1}^2}{\sigma_{M_0}^2 + \tau_{M_0}^2}. \tag{3.5}$$

This can be implemented in *R* using *Mod1* as follows:

```
# Get the within-/between-cluster variance from Mod1
estVar1   = VarCorr(Mod1)
tausq1    = as.numeric(estVar1[1,1])
sigmasq1  = as.numeric(estVar1[2,1])
# Then estimate R-square
estRsq1 = 1- (sigmasq1+tausq1)/(sigmasq0+tausq0)
# print the R-square value at level 1
estRsq1
```

```
## [1] 0.1840543
```

Results indicate that level 1 of *Mod1* explains approximately 18.4% of the variance in the *Written* score above and beyond the variance accounted for in the null model *Mod0*.

And for level 2, R^2 can be calculated as

$$R_2^2 = 1 - \frac{\sigma_{M_1}^2/B + \tau_{M_1}^2}{\sigma_{M_0}^2/B + \tau_{M_0}^2}, \tag{3.6}$$

where B is the average size of the level 2 units (e.g., schools).

We calculate R^2 as follows:

```
# We first calculate the number of individuals in the sample
N = dim(dSci)[1]; N
```

```
## [1] 1905
```

```
# Then we calculate the number of schools
J = length(unique(dSci$Centre)); J
```

```
## [1] 73
```

```
# Now we can calculate the average size
B = N/J; B
```

```
## [1] 26.09589
```

```
# Then the estimate R-Square at level 2
estRsq2 = 1- (sigmasq1/B+tausq1)/(sigmasq0/B+tausq0)
# print the value
estRsq2
```

```
## [1] 0.1392446
```

This means that the *Centre* (classes) level (i.e., level 2) explains 13.9% of variations above and beyond the variance explained by level 1.

3.2.6   If Classical Linear Regression Is Used

As stated previously, when using a linear regression to analyze nested data without accounting for the nested structure, we expect an erroneous smaller variance estimate. Let us explore this assumption:

```
# Regression
slm1 = lm(Course~Written, data = dSci)
summary(slm1)
```

```
##
## Call:
## lm(formula = Course ~ Written, data = dSci)
##
## Residuals:
##      Min      1Q  Median      3Q     Max
## -69.498  -9.083   1.747  11.370  39.591
##
## Coefficients:
##              Estimate Std. Error t value Pr(>|t|)
## (Intercept) 50.28311    1.31010   38.38   <2e-16 ***
## Written      0.38366    0.01681   22.82   <2e-16 ***
## ---
## Signif. codes:  0 '***' 0.001 '**' 0.01 '*' 0.05 '.' 0.1 ' ' 1
##
## Residual standard error: 15.72 on 1903 degrees of freedom
## Multiple R-squared:  0.2149, Adjusted R-squared:  0.2145
## F-statistic:    521 on 1 and 1903 DF,  p-value: < 2.2e-16
```

It can be seen that the standard errors for the parameters β_0 and β_1 are underestimated from 1.7195 and 0.0172 in *Mod1* to 1.3101 and 0.0168 in *slm1*. The *t*-values are larger and increased from 29.081 and 21.997 in *Mod1* to 38.38 and 22.82 in *slm1*.

3.2.7 Fitting MLM with Random-Slope Model

In the random-intercept model above, we assume the impact of the independent variable on the dependent variable is the same for all the schools in the second level, and this assumption is not always true. Instead, let us allow the impact of the independent variable on the dependent variable to vary across the level 2 effects. In the context of the current research problem, it means that we allow the impact of *Written* on *Course* to vary from one school to another.

In this case, we come to the general random-intercept and random-slope model, which is described in model (3.1). In this model, $\beta_0 + \beta_1 x_{ij}$ is the fixed-effects component and $U_{0j} + U_{1j} x_{ij} + \epsilon_{ij}$ is the random-effects component. In this general two-level MLM, we estimate the overall parameters for the intercept β_0 and the slope β_1, along with the within-cluster standard deviation σ, the between-cluster standard deviations for the random-intercept τ_0 and random-slope τ_1 as well as the correlation between the random-intercept and random-slope ρ_{01}.

For *R* implementation, in the random-intercept model above, 1 stood for the random-intercept (i.e., *random = ~1|Centre*). However, if we want the level 1 slope to vary randomly, we will change the syntax to: *random = ~Written|Centre* as follows:

```
# Fitting a random-slope model
Mod2 = lme(fixed = Course~Written, random =~Written|Centre,
           data = dSci)
# Print model summary
summary(Mod2)
```

```
## Linear mixed-effects model fit by REML
##  Data: dSci
##        AIC      BIC    logLik
##   15450.32 15483.62 -7719.158
##
## Random effects:
##  Formula: ~Written | Centre
##  Structure: General positive-definite, Log-Cholesky parametrization
##             StdDev      Corr
## (Intercept) 15.8373690 (Intr)
## Written      0.1522478 -0.837
## Residual    13.1572085
##
## Fixed effects: Course ~ Written
##              Value Std.Error   DF  t-value p-value
## (Intercept) 52.29130 2.4926220 1831 20.97843       0
## Written      0.35769 0.0271809 1831 13.15960       0
##  Correlation:
```

```
##            (Intr)
## Written -0.895
##
## Standardized Within-Group Residuals:
##          Min         Q1        Med         Q3        Max
## -5.4218287 -0.5191729  0.1077014  0.6475218  2.8984426
##
## Number of Observations: 1905
## Number of Groups: 73
```

From the summary in *Mod2*, we can see the following.

- *Better Model Fitting*:

Compared to the random-intercept model, the AIC, BIC, and -logLik statistics dropped further from 15476.84, 15499.04, and 7734.418 in the random-intercept model (*Mod1*) to 15450.32, 15483.62, and 7719.158 in *Mod2*.

- *Parameter Estimates*:

From the *fixed-effects* component, we can see that *Written* is still a significant predictor of *Course* ($t = 13.159, p < 0.0001$). As *Written* score increases by 1 point, the score of the *Course* increases by 0.358 points.

3.2.8 Fitting Interactions–Moderation Analysis

Interactions (or moderation) among the predictor variables can be very important in the application of multi-level models.

To create an interaction term within MLM, we can multiply the predictors. In this example, we would like to examine the relationship between the score of the course (i.e., *Course*) and the score of the written paper (i.e., *Written*) among *Gender* (0 = boy and 1 = girl) to see whether or not there is a gender difference and whether *Gender* moderates the relationship between *Written* and *Course*. For this purpose, we can create an interaction involving *Written* and *Gender* from the random-slope model (*Mod2*) as follows:

```
# Fitting a random-slope MLM with interaction
Mod3 = lme(fixed = Course~Written*Gender,
random = ~Written|Centre, data =dSci)
# Print the summary
summary(Mod3)
```

```
## Linear mixed-effects model fit by REML
##   Data: dSci
##          AIC       BIC    logLik
##    15254.02  15298.42  -7619.011
##
## Random effects:
##   Formula: ~Written | Centre
##    Structure: General positive-definite, Log-Cholesky parametrization
```

```
##             StdDev     Corr
## (Intercept) 15.7494068 (Intr)
## Written      0.1449233 -0.842
## Residual    12.4568839
##
## Fixed effects: Course ~ Written * Gender
##                   Value Std.Error   DF  t-value p-value
## (Intercept)    43.47724 2.8302948 1829 15.361381  0.0000
## Written         0.40405 0.0312758 1829 12.919043  0.0000
## Gender         11.43055 2.3047432 1829  4.959575  0.0000
## Written:Gender -0.03314 0.0293337 1829 -1.129887  0.2587
##  Correlation:
##                (Intr) Writtn Gender
## Written        -0.914
## Gender         -0.510  0.556
## Written:Gender  0.481 -0.562 -0.965
##
## Standardized Within-Group Residuals:
##         Min          Q1         Med          Q3         Max
## -5.31147203 -0.51378197  0.09307683  0.64196445  2.80795340
##
## Number of Observations: 1905
## Number of Groups: 73
```

In *Mod3*, the interaction term *Written:Gender* is not statistically significant with *p-value = 0.2587*. With this non-significant interaction, we can fit a reduced main effect model as follows:

```
# Fitting a random-slope MLM with main effect
Mod4 = lme(fixed = Course~Written+Gender,
random = ~Written|Centre, data =dSci)
# Print the summary
summary(Mod4)
```

```
## Linear mixed-effects model fit by REML
##  Data: dSci
##        AIC      BIC    logLik
##   15248.07 15286.93 -7617.036
##
## Random effects:
##  Formula: ~Written | Centre
##  Structure: General positive-definite, Log-Cholesky parametrization
##             StdDev     Corr
## (Intercept) 15.7347859 (Intr)
## Written      0.1437653 -0.843
## Residual    12.4605921
##
## Fixed effects: Course ~ Written + Gender
##                Value Std.Error   DF  t-value p-value
## (Intercept) 44.99983 2.4781803 1830 18.15842       0
## Written      0.38437 0.0257669 1830 14.91729       0
## Gender       8.92064 0.6084613 1830 14.66099       0
##  Correlation:
##         (Intr) Writtn
## Written -0.888
## Gender  -0.198  0.068
##
```

```
## Standardized Within-Group Residuals:
##          Min            Q1          Med            Q3          Max
## -5.27518826  -0.51746605   0.09295705   0.64202009   2.82067353
##
## Number of Observations: 1905
## Number of Groups: 73
```

Mod4 results indicate that the estimated intercept (i.e., the score of *Course* when *Written* is zero) is 44.999 for boy (i.e., 0 = boy) and 44.999 + 8.921 = 53.92 for girl (i.e., 1 = girl) with the same slope of change for both gender at 0.384. This also indicates that the score of the *Course* is significantly higher for girls than boys with the difference of 8.921.

This can be graphically illustrated with *R panel* plot to *xyplot* to add a regression line to Fig. 3.1 as follows:

```
# Add panel to xyplot
xyplot(Course~Written|as.character(Gender),group=Centre,
data=dSci,type=c("p","r"),
        xlab="The Score of Written Paper",ylab="The Score of the Course",
        panel = function(x,y,...){
        panel.xyplot(x,y,...)
        panel.abline(lm(y~x),lwd=4, col="red")
        panel.abline(h=coef(lm(y~x))[1],lwd=3, lty=4,col="blue")
})
```

As seen from Fig. 3.4, the overall slopes (i.e., the thick red lines) for both *girls* and *boys* are very similar, but their intercepts (i.e., the dashed blue horizontal lines) are different with the intercept for *girls* (i.e., *Gender = 1*) is 8.921 units higher than that for *boys* (i.e., *Gender = 0*).

3.3   Model Selection with Likelihood Ratio Chi-Square Test

3.3.1   Likelihood Ratio Chi-Square Test

We previously showed how AIC and BIC can be used to compare model fit. However, these statistics are descriptive in nature, and there is no formal hypothesis test to compare different AIC and BIC values about which model is better can be made. For example, if the AIC for one model is 300 and 310 for another model, we cannot know if the difference of 10 units in AIC is statistically significant, which can be generalizable from the sample to the general population.

However, hypothesis testing is possible using the likelihood ratio Chi-Square difference test with the *logLik* value in the model fitting. This test is based on the deviance statistic formulated from the likelihood function, and it allows us to test to determine if the overall fit of two models is different. For model selection under the principle of model parsimony, the more constrained model (i.e., simpler model or more parsimonious model) should be selected if no differences exist. However, if a significant difference is found, the model with the better model fit should be used.

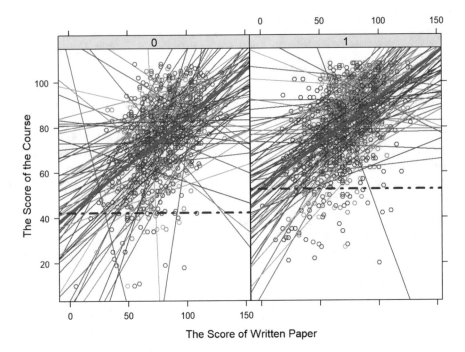

Fig. 3.4 The score of the course vs. the score of written paper by gender

The difference in Chi-Square values can be obtained using the *anova()* function command. Note that when using the *nlme* package, the *anova()* command will provide accurate comparisons only if maximum likelihood estimation is used. When maximum likelihood is used, both the fixed and random effects are compared simultaneously. When restricted maximum likelihood (i.e., *REML*) is used, only the random effects are compared. Since the default method in *lme* is *REML*, we need to refit the models above using the *ML* option.

3.3.2 Refit the Models with Maximum Likelihood Estimation

We can make *R* function *update* to update the previous models with the option *method="ML"* as follows:

```
# refit the null model using ML
Mod0.2 =update(Mod0, method="ML")
summary(Mod0.2)

## Linear mixed-effects model fit by maximum likelihood
##  Data: dSci
##         AIC      BIC      logLik
```

```
##    15901.91 15918.57 -7947.954
##
## Random effects:
##  Formula: ~1 | Centre
##          (Intercept) Residual
## StdDev:     9.433534 15.05465
##
## Fixed effects: Course ~ 1
##                 Value Std.Error   DF  t-value p-value
## (Intercept) 79.35968  1.197539 1832 66.26898       0
##
## Standardized Within-Group Residuals:
##        Min         Q1        Med        Q3        Max
## -4.9459949 -0.5191447  0.1138137  0.6820928  2.6976359
##
## Number of Observations: 1905
## Number of Groups: 73
```

```
# Refit the Random-Intercept MLM
Mod1.2 =update(Mod1, method="ML")
summary(Mod1.2)
```

```
## Linear mixed-effects model fit by maximum likelihood
##  Data: dSci
##        AIC      BIC    logLik
##    15472.6 15494.8 -7732.298
##
## Random effects:
##  Formula: ~1 | Centre
##          (Intercept) Residual
## StdDev:     8.782584 13.42432
##
## Fixed effects: Course ~ Written
##                 Value Std.Error   DF  t-value p-value
## (Intercept) 50.30195 1.7245711 1831 29.16780       0
## Written      0.37784 0.0171785 1831 21.99472       0
##  Correlation:
##         (Intr)
## Written -0.766
##
## Standardized Within-Group Residuals:
##        Min         Q1        Med        Q3        Max
## -5.2524761 -0.5362972  0.1013296  0.6767829  2.9183640
##
## Number of Observations: 1905
## Number of Groups: 73
```

```
# Refit the Random-Slope MLM
Mod2.2 =update(Mod2, method="ML")
summary(Mod2.2)
```

```
## Linear mixed-effects model fit by maximum likelihood
##  Data: dSci
##        AIC        BIC     logLik
##   15446.98 15480.29 -7717.49
##
## Random effects:
##  Formula: ~Written | Centre
##  Structure: General positive-definite, Log-Cholesky parametrization
##             StdDev       Corr
## (Intercept) 15.6424732 (Intr)
## Written      0.1496741 -0.836
## Residual    13.1576786
##
## Fixed effects: Course ~ Written
##                 Value Std.Error   DF  t-value p-value
## (Intercept) 52.27346 2.4731084 1831 21.13674       0
## Written      0.35788 0.0269429 1831 13.28280       0
##  Correlation:
##          (Intr)
## Written -0.895
##
## Standardized Within-Group Residuals:
##        Min         Q1        Med         Q3        Max
## -5.4204143 -0.5185483  0.1072659  0.6468811  2.8992589
##
## Number of Observations: 1905
## Number of Groups: 73
```

```
# Refit the Gender-Interaction MLM
Mod3.2 =update(Mod3, method="ML")
summary(Mod3.2)
```

```
## Linear mixed-effects model fit by maximum likelihood
##  Data: dSci
##        AIC        BIC     logLik
##   15246.16 15290.58 -7615.082
##
## Random effects:
##  Formula: ~Written | Centre
##  Structure: General positive-definite, Log-Cholesky parametrization
##             StdDev      Corr
## (Intercept) 15.550525 (Intr)
## Written      0.142346 -0.841
## Residual    12.450629
##
## Fixed effects: Course ~ Written * Gender
##                     Value Std.Error   DF  t-value p-value
## (Intercept)     43.46613 2.8138761 1829 15.447067  0.0000
## Written          0.40413 0.0310852 1829 13.000751  0.0000
## Gender          11.41894 2.3047212 1829  4.954587  0.0000
## Written:Gender  -0.03296 0.0293310 1829 -1.123830  0.2612
##  Correlation:
##                 (Intr) Writtn Gender
## Written         -0.914
## Gender          -0.513  0.560
## Written:Gender   0.484 -0.565 -0.965
##
## Standardized Within-Group Residuals:
```

```
##              Min              Q1           Med            Q3            Max
## -5.31276852 -0.51404456    0.09295555    0.64176733    2.81030136
##
## Number of Observations: 1905
## Number of Groups: 73
```

```
# Refit the Gender Main Effects MLM
Mod4.2 =update(Mod4, method="ML")
summary(Mod4.2)
```

```
## Linear mixed-effects model fit by maximum likelihood
##  Data: dSci
##        AIC       BIC     logLik
##    15245.42  15284.29  -7615.712
##
## Random effects:
##  Formula: ~Written | Centre
##  Structure: General positive-definite, Log-Cholesky parametrization
##              StdDev      Corr
## (Intercept) 15.5396939  (Intr)
## Written      0.1412809  -0.842
## Residual    12.4576261
##
## Fixed effects: Course ~ Written + Gender
##                Value Std.Error    DF  t-value p-value
## (Intercept) 44.98135 2.4592667  1830 18.29056       0
## Written      0.38455 0.0255409  1830 15.05631       0
## Gender       8.92257 0.6086747  1830 14.65901       0
##  Correlation:
##          (Intr) Writtn
## Written  -0.887
## Gender   -0.200   0.069
##
## Standardized Within-Group Residuals:
##          Min              Q1           Med            Q3            Max
## -5.27528487 -0.51760193    0.09220942    0.64291247    2.82219774
##
## Number of Observations: 1905
## Number of Groups: 73
```

3.3.3   ANOVA for Model Selection

Now, we can call *anova* for model comparison and selection. The *anova* function in *R* is used to compute *analysis of variance* and *analysis of deviance* tables for one or more fitted model objects. When given a sequence of model objects, *anova* provides statistical significance tests to the models against one another in the order specified (Chambers and Hastie 1992).

To compare the null model *Mod0* with the random-intercept model *Mod1*, the following *R* code chunk can be called:

```
# model comparison
anova(Mod0.2, Mod1.2)
```

```
##         Model df    AIC      BIC    logLik  Test  L.Ratio p-value
## Mod0.2      1  3 15901.91 15918.57 -7947.954
## Mod1.2      2  4 15472.59 15494.80 -7732.298 1 vs 2 431.3129 <.0001
```

The likelihood ratio statistic is 431.313 and the *p-value* < *0.0001* from the χ^2-test. This indicates that the added covariate of *Written* is a statistically significant predictor to the score of the *Course* and substantially improved the model fit. Among these two models, the random-intercept MLM *Mod1.2* is better than the null MLM *Mod0.2*.

Similarly, we can compare the random-slope model *Mod2.2* to the random-intercept only model *Mod1.2* as follows:

```
# model comparison
anova(Mod1.2, Mod2.2)
```

```
##         Model df    AIC      BIC    logLik  Test  L.Ratio p-value
## Mod1.2      1  4 15472.59 15494.80 -7732.298
## Mod2.2      2  6 15446.98 15480.29 -7717.490 1 vs 2 29.61557 <.0001
```

Again, this likelihood ratio χ^2-test gives a significant model comparison (i.e., *p-value* < *0.0001*), which indicates that the random-slope model *Mod2.2* is more preferred to fit this data. This finding is consistent with the graphical presentation in Fig. 3.1, where the slopes are quite different from each school.

We can also use *anova* to assess whether *Gender* is a significant moderator that modifies the effect of *Written* on *Course* statistically. This can be implemented to compare the *Mod2.2* to *Mod4.2* as follows:

```
# model comparison
anova(Mod2.2, Mod4.2)
```

```
##         Model df    AIC      BIC    logLik  Test  L.Ratio p-value
## Mod2.2      1  6 15446.98 15480.29 -7717.490
## Mod4.2      2  7 15245.42 15284.29 -7615.712 1 vs 2 203.5559 <.0001
```

Again, the likelihood ratio χ^2-test gives a significant model comparison (i.e., *p-value* < *0.0001*), which indicates that the random-slope model with *Gender* (i.e., *Mod4.2*) is statistically significantly better than the random-slope model *Mod2.2*, and it is preferred to fit this data. With *Mod4.2*, the *Gender* is a significant predictor in modeling the score of the *Course*.

3.4 Confidence Intervals for Parameter Estimates

The default output for *nlme* only provides statistical significance tests for the parameter estimates in the fixed-effects part, and there is no such statistical significance for the variances in random-effects part. However, the statistical significance of random effects can provide useful information about between-cluster and within-cluster variabilities. For example, the significance of the random-intercept would indicate that there exists variation in the score of *Course* among schools in the data. Different schools exhibit significantly different mean scores of *Course*. Similarly, a significant random-slope (e.g., *Written*) indicates significant variation in the impact of the score of *Written* paper on the score of the *Course* across the schools. This information provides useful insights into the factors that contribute to the differences in score of *Course* among schools and would validate further modeling on other factors. Unfortunately, the default setting in *nlme* does not provide an option for testing the significance of random effects.

However, information about significance of random effects can be obtained by creating confidence intervals. With the *nlme* package, the function *intervals()* can provide 95% confidence intervals for both the fixed effects and the variances of random effects. The variances of the random-effects confidence intervals can be used to determine the significance of the random effects.

Using the random-effects model *Mod4*, we could conclude that both the scores of *Written* paper and *Gender* are significant predictors of the score of the *Course*. However, we could not determine whether random-intercept or random-slope variability significantly differed from 0. If not different, the result would indicate that there is no statistically significant variations in the mean score of the *Course* (i.e., variation in random-intercept), and the relationship between the score of the *Course* and the score of *Written* paper across schools (i.e., variation in random-slope). In this situation, we would not need the random-intercept and random-slope MLM.

To determine their significance, we can use the *intervals()* as follows:

```
intervals(Mod4)
```

```
## Approximate 95% confidence intervals
##
##   Fixed effects:
##                     lower        est.       upper
## (Intercept) 40.139469 44.9998274 49.8601862
## Written      0.333837  0.3843727  0.4349083
## Gender       7.727293  8.9206443 10.1139958
## attr(,"label")
## [1] "Fixed effects:"
##
##   Random Effects:
##     Level: Centre
##                        lower        est.       upper
## sd((Intercept))   12.0706513 15.7347859 20.5111955
```

```
## sd(Written)                          0.1016228   0.1437653   0.2033841
## cor((Intercept),Written)  -0.9224197  -0.8428226  -0.6945355
##
##  Within-group standard error:
##     lower      est.      upper
## 12.05678 12.46059 12.87793
```

As seen from the output, the estimated standard deviation for the random-intercept τ_0 is $\hat{\tau}_0 = 15.735$ and the 95% confidence interval is (12.063, 20.524), which is statistically significant. Likewise, the 95% confidence interval for the random-slope variance is (0.102, 0.204) with $\hat{\tau}_1 = 0.144$, which is also statistically significant.

3.5 Exercises

Using the *Health Behavior in School-Aged Children (HBSC)* dataset in Appendix A.3,

1. perform a two-level MLM with *Student* (named as *CASEID*) as the level 1 and *School* (named as *SCHL_ID*) as level 2 to test whether *Physical Activity* (named as *physact*) is significantly related to *Body Mass Index* (named as *BMI*);
2. test whether this relationship between *Physical Activity* (named as *physact*) and *Body Mass Index* (named as *BMI*) is different by *Student's sex* (named as *sex*) while accounting for the two-level nested structure.

Chapter 4
Higher-Level Multi-Level Modeling

Going beyond the two-level MLM from Chap. 3, we discuss higher-level MLM in this chapter with a focus on three-level MLM. The *R* package *nlme* (Pinheiro and Bates 2000) can easily be used to fit MLM with three or more levels by changing the option *random* in the function *lme*. We will use the data on *Performance for Pupils in 49 Inner London Schools* in Sect. A.4.

Note to readers: We will use *R* package *nlme* (i.e., *nonlinear mixed-effects* model) in this chapter. Remember to install this *R* package to your computer using *install.packages("nlme")* and load this package into *R* session using *library(nlme)* before running all the *R* programs in this chapter.

4.1 Data on Performance for Pupils in 49 Inner London Schools

4.1.1 Data Description

In this data, there are over 1192 students from 49 schools (numbered from 1 to 50, but school with $ID = 43$ is missing) measured over three junior school years with 3236 observations included in this dataset. The nested data structure is three-leveled, where the 1192 students are nested within classrooms and the classrooms are nested within the 49 schools. In the 49 schools, a few of them have 4 classrooms, but most have only 1 or 2 classrooms. Students' performance is tracked from their three junior school years and is coded as *SchoolYear*, with year one $= 0$, year two $= 1$, and year three $= 2$. We can also consider these three longitudinal observations as another level. In that case, we would have a 4-level nested data structure. Since we will discuss longitudinal data analysis in a separate chapter, we will only focus on three-level MLM in this chapter.

© The Author(s), under exclusive license to Springer Nature Switzerland AG 2021
D.-G. (Din) Chen, J. K. Chen, *Statistical Regression Modeling with R*,
Emerging Topics in Statistics and Biostatistics,
https://doi.org/10.1007/978-3-030-67583-7_4

The dataset can be loaded into *R* as follows:

```
# Read in data#
dJSP = read.csv("JSP.csv", header=T)
# Check the dimension of the data
dim(dJSP)
```

```
## [1] 3236      9
```

```
# Check how many schools are in this data
length(unique(dJSP$School))
```

```
## [1] 49
```

```
# Check how many students are in this data
length(unique(dJSP$ID))
```

```
## [1] 1192
```

```
# Check how many classes are in this data
length(unique(dJSP$Class))
```

```
## [1] 4
```

To display the nested structure in the data, we can use *R* function *xtabs* with format function *ftable* as follows:

```
# Check the nested data structure and print the data for 10 Schools
head(ftable(xtabs(~School+Class, dJSP)),10)
```

```
##
##                    "Class" "1" "2" "3" "4"
##    "School"
##    "1"                     45  47   0   0
##    "2"                     34   2   0   0
##    "3"                     41   7   0   0
##    "4"                     72   0   0   0
##    "5"                     39  44   0   0
##    "6"                     36  24   0   0
##    "7"                     25  14   0   0
##    "8"                     33  37  17   0
##    "9"                     24  35   6   0
##    "10"                    24   0   0   0
```

As we have known that there are 49 schools (i.e., *School*) in this dataset, we only printed the nesting structure for the first 10 schools with their classrooms (i.e., *Class*) using the *R* function *head*. As noted from the above output, there are two classrooms (i.e., "1" and "2") nested within school 1 where 45 and 47 observations (due to several *SchoolYear*) are from classrooms 1 and 2, respectively. Similarly, there are two classrooms nested within other five schools (i.e., 2, 3, 5, 6, and 7), only one classroom within two schools (i.e., 4 and 10), and three classrooms within two schools (i.e., 8 and 9).

To illustrate three-level MLM, we make use of the data from the first year of junior high students, and we create a subset of the data as follows:

```
# Subset to get the first year's data
d1 = dJSP[dJSP$SchoolYear==0,]
# Show their structure
head(ftable(xtabs(~School+Class, d1)),10)
```

```
##
##                "Class" "1" "2" "3" "4"
##   "School"
##   "1"                    17  17   0   0
##   "2"                    11   1   0   0
##   "3"                    17   3   0   0
##   "4"                    24   0   0   0
##   "5"                    14  14   0   0
##   "6"                    12   8   0   0
##   "7"                     9   5   0   0
##   "8"                    11  13   6   0
##   "9"                     8  12   2   0
##   "10"                   13   0   0   0
```

As noted from the above output for the dataset *d1*, there are different numbers of students (i.e., student *ID* as level 1) nested within classrooms (i.e., *Class* as level 2) and different numbers of classrooms (maximum of 4 classes) nested within these 49 schools (i.e., *School* as level 3). For example, there are 17 students each for two classrooms (i.e., *Class* 1 and 2) nested within the first *School* (i.e., "1"). In this data, *nesting* implies that all students within a classroom attend the same school.

4.1.2   Graphical Analysis

With this data, there are several variables that can be used to illustrate three-level MLM. We will use the English test score (i.e., *English*) as the outcome variable and see whether or not the Raven's Progressive Matrices in year 1 (i.e., *RavensTest1*) is correlated with their *English* scores. For those unfamiliar with Raven's Progressive Matrices, it is non-verbal group tests that are typically used in educational settings. They are used to measure abstract reasoning and are regarded as a non-verbal estimate of fluid intelligence.

Let us graphically illustrate the relationship between the *English* score and *RavensTest1* to see whether a linear model can be reasonably assumed for each

Class nested within *School*. For this purpose, we can use the *R* function *xyplot* from the library *lattice* in the following *R* code chunk:

```
par(mar = c(4, 4, .1, .1))
# Load the lattice library
library(lattice)
# Call xyplot to make the plot
xyplot(English~RavensTest1|as.factor(School),
group=Class, type=c("p","r"), d1)
```

In Fig. 4.1, we plotted the data (points) with a fitted regression line for each classroom nested within schools. As seen from this figure, there are different numbers of regression lines for the 49 schools (school 43 is missing) since there are different number of classes nested within different schools.

It can be seen from this figure that an approximate positive linear relationship exists between the *RavensTest1* and the *English* scores. Generally, the higher the score of *RavensTest1*, the higher the score of *English* test. Due to the nested data structure, a linear MLM should be used to better analyze this data.

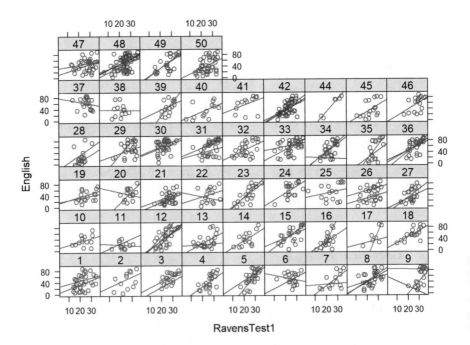

Fig. 4.1 Relationship between Raven's score and English score for all students

4.2 Understanding Three-Level MLMs

In three-level MLMs, we specify that level 2 (denoted by j) is nested in level 3 (denoted by k) as follows:

$$(\text{Level-1}) \quad y_{ijk} = \beta_{0jk} + \beta_{1jk} x_{ijk} + \epsilon_{ijk}, \tag{4.1}$$

where i represents the level 1 that is nested in level 2 (denoted by j). And k represents the level 3 cluster to which the level 2 (i.e., j) belongs. In our example, i represents the students (i.e., level 1) who are nested within classrooms (i.e., level 2, j), and the classrooms (j) are nested within schools (i.e., level 3, k).

In model 4.1, y_{ijk} and x_{ijk} represent the dependent and independent variables. The β_{0jk} and β_{1jk} are regression parameters of intercepts and slopes, which are level 2 (j) and level 3 (k) specific. ϵ_{ijk} represents the within-cluster error term and is assumed to be normally distributed with a mean of 0 and a variance of σ^2.

To formulate three-level MLM, those intercepts and slopes should be chosen to be specified either as random effects at both levels 2 and 3 or at only one level. Their specifications are always based on theory and empirical evidence from research questions. Typically, the most general specification assumes that these intercepts and slopes are random at both levels 2 and 3. With this assumption, we can then simplify it to the most parsimonious model with statistical testing. When these coefficients are not random at both levels, related randomness terms are removed from subsequent models as a part of the model selection process.

The general specification for random-intercept and random-slope in both levels 2 and 3 is as follows:

Level 2 :

$$\beta_{0jk} = \beta_{00k} + U_{0jk}$$
$$\beta_{1jk} = \beta_{10k} + U_{1jk}$$

Level 3 :

$$\beta_{00k} = \beta_0 + V_{00k}$$
$$\beta_{10k} = \beta_1 + V_{10k}, \tag{4.2}$$

where β_0 and β_1 are the overall intercept and slope parameters to be estimated.

In Eq. 4.2, V_{00k} and V_{10k} are the level 3 random effects for the intercept and slope that are assumed as bivariate normal distributions with error distributions as follows:

$$\begin{pmatrix} V_{00k} \\ V_{10k} \end{pmatrix} \sim N \left(\begin{pmatrix} 0 \\ 0 \end{pmatrix}, \begin{pmatrix} \tau_{0.3}^2 & \tau_{0.3}\tau_{1.3}\rho_{01.3} \\ \tau_{0.3}\tau_{1.3}\rho_{01.3} & \tau_{1.3}^2 \end{pmatrix} \right),$$

where $\tau_{0.3}$, $\tau_{1.3}$, and $\rho_{01.3}$ are the level 3 between-cluster standard deviation for intercept, the standard deviation for slope, and the correlation.

In Eq. (4.2), U_{0jk} and U_{1jk} are the level 2 random effects for the intercept and slope and are assumed as bivariate normal distributions with error distributions as follows:

$$\begin{pmatrix} U_{0jk} \\ U_{1jk} \end{pmatrix} \sim N\left(\begin{pmatrix} 0 \\ 0 \end{pmatrix}, \begin{pmatrix} \tau_{0.2}^2 & \tau_{0.2}\tau_{1.2}\rho_{01.2} \\ \tau_{0.2}\tau_{1.2}\rho_{01.2} & \tau_{1.2}^2 \end{pmatrix} \right),$$

where $\tau_{0.2}$, $\tau_{1.2}$, and $\rho_{01.2}$ are the level 2 between-cluster standard deviation for the intercept, the standard deviation for the slope, and the correlation, respectively.

Substituting the two level 3 equations in model 4.2 to the level 2 equations, we can get the three-level random-intercept and random-slope as follows:

$$\beta_{0jk} = \beta_0 + V_{00k} + U_{0jk}$$
$$\beta_{1jk} = \beta_1 + V_{10k} + U_{1jk}. \tag{4.3}$$

Furthermore, these terms may be substituted into Eq. 4.1 to provide the full three-level MLM as follows:

$$y_{ijk} = \beta_0 + V_{00k} + U_{0jk} + (\beta_1 + V_{10k} + U_{1jk})x_{ijk} + \epsilon_{ijk}$$
$$= \beta_0 + \beta_1 x_{ijk} + V_{00k} + U_{0jk} + (V_{10k} + U_{1jk})x_{ijk} + \epsilon_{ijk}, \tag{4.4}$$

where $\beta_0 + \beta_1 x_{ijk}$ is the fixed effects and $V_{00k} + U_{0jk} + (V_{10k} + U_{1jk})x_{ijk} + \epsilon_{ijk}$ is the random effects. Together, this model is called mixed-effects model. In this mixed-effects model, we estimated the fixed-effects regression parameters β_0 and β_1 similar to linear regression. We also estimate the parameters associated with the random effects that include the within-cluster variance σ^2 (similar to linear regression), the level 2 between-cluster $\tau_{0.2}$, $\tau_{1.2}$, and $\rho_{01.2}$, and the level 3 between-cluster $\tau_{0.3}$, $\tau_{1.3}$, and $\rho_{01.3}$.

4.3   R Implementation with *lme*

The syntax for the three-level and two-level MLMs is very similar, except we define an extra level (i.e., level 3) in R. To define a model with more than two levels, we must do two things:

- include the variables denoting the nesting structures: *School* (school-level influence) and *Class* (classroom-level influence); and
- designate the nesting structure of the levels (*Students* nested within *Classrooms* within *Schools*). The nesting structure in *lme* is defined as *A/B* where *A* is the higher-level data unit (e.g., *School*) and *B* is the lower unit (e.g., *Classroom*).

4.3.1   Fitting the Null Model

First, we define a null model to estimate the mean effect of the student's score of
Course without any other predictors. The general three-level MLM in (4.4) is then
simplified as the three-level null model equation as follows:

$$y_{ijk} = \beta_0 + V_{00k} + U_{0jk} + \epsilon_{ijk}. \tag{4.5}$$

In this null model, we estimate the fixed-effects overall intercept parameter, β_0,
and the three random effects of within-cluster variance of σ^2 and between-cluster
variances $\tau_{0.2}^2$ and $\tau_{0.3}^2$ in the random effects $V_{00k} + U_{0jk} + \epsilon_{ijk}$.

The syntax to fit a 3-level null model appears below. Model results are stored in
the object *Mod3.0*.

```
# Fit the null model
# Load the package
library(nlme)
# Fit the null model
Mod3.0 = lme(fixed=English~1, random=~1|School/Class, data = d1)
# Print the summary
summary(Mod3.0)
```

```
## Linear mixed-effects model fit by REML
##   Data: d1
##        AIC      BIC    logLik
##    10598.31 10618.51 -5295.154
##
## Random effects:
##  Formula: ~1 | School
##         (Intercept)
## StdDev:    5.562036
##
##  Formula: ~1 | Class %in% School
##         (Intercept) Residual
## StdDev:     7.91796 22.81719
##
## Fixed effects: English ~ 1
##                 Value Std.Error   DF  t-value p-value
## (Intercept) 47.42434  1.379502 1061 34.37787       0
##
## Standardized Within-Group Residuals:
##         Min          Q1         Med          Q3         Max
## -2.29293914 -0.81549675 -0.02260965  0.81467574  2.19968499
##
## Number of Observations: 1154
## Number of Groups:
##              School Class %in% School
##                  49                93
```

As seen from the above model fitting, we can extract:

• The overall intercept estimate:

The overall intercept is presented in *Fixed effects: English ~ 1*, which is estimated with $\hat{\beta}_0 = 47.424$, SE $= 1.379$, t-value $= 34.378$, and p-value $= 0$. This means that the overall score of this course is estimated at 47.424 for all students and is statistically significantly different from zero.

• Level 3 random effects, *School*:

The level 3 random effects for intercept are presented in *Formula: ~1|School*, which are estimated as $\hat{\tau}_{0,3} = 5.562$. This means that in this data the intercepts vary across schools with a standard deviation of 5.562.

We can also extract this number from the estimated covariance matrix as follows:

```
# Get the covariance matrix from all random-effects estimates
estRand = VarCorr(Mod3.0)
# show all
estRand
```

```
##                    Variance        StdDev
## School =           pdLogChol(1)
## (Intercept)        30.93625          5.562036
## Class =            pdLogChol(1)
## (Intercept)        62.69410          7.917960
## Residual           520.62399        22.817186
```

```
# Now extract the variance random-effects for level-3 of "school":
est.tausq03 = estRand[2,1];est.tausq03
```

```
## [1] " 30.93625"
```

```
# The standard deviation
est.tau03=estRand[2,2];est.tau03
```

```
## [1] " 5.562036"
```

• Level 2 random effects: *Class nested in School*:

The level 2 standard deviation estimate for the intercept is presented in *Formula: ~1 | Class %in% School*, which is $\hat{\tau}_{0,2} = 7.918$. This means that the intercepts vary across classrooms within schools with a standard deviation of 7.918. This random-effect estimate can be extracted as follows:

```
# The between-cluster variance for intercept
est.tausq02= estRand[4,1];est.tausq02
```

```
## [1] " 62.69410"
```

```
# The between-cluster standard deviation
est.tau02 = estRand[4,2];est.tau02
```

   ## [1] " 7.917960"

- Level 1 variance estimate: σ^2
   The within-cluster, i.e., individual-level, variance can be accessed as follows:

```
# The within-cluster variance
est.sigmasq = estRand[5,1];est.sigmasq
```

   ## [1] "520.62399"

```
# The estimated within-cluster standard deviation
est.sigma = estRand[5,2];est.sigma
```

   ## [1] "22.817186"

- Data structure and sample size:
   The model output also presents the sample size for each of the higher-level units. This information helps us make sure that the model is properly defined and the correct data is being used. In this analysis, we can see the *Number of Observations: 1154*. And in the *Number of Groups*, we can see that the number of *Schools* is 49 and all *Class %in% School* is 93, which confirms what we know about this dataset.

- Confidence intervals for all parameters:

Finally, we can use the *intervals* function to obtain confidence intervals for estimated parameters:

```
intervals(Mod3.0)
```

```
## Approximate 95% confidence intervals
##
##   Fixed effects:
##                    lower      est.     upper
## (Intercept) 44.71748 47.42434 50.1312
## attr(,"label")
## [1] "Fixed effects:"
##
##   Random Effects:
##     Level: School
##                       lower      est.      upper
## sd((Intercept)) 2.729494 5.562036 11.33406
##     Level: Class
##                       lower      est.      upper
## sd((Intercept)) 5.331497 7.91796 11.75919
```

```
##
##  Within-group standard error:
##    lower      est.      upper
## 21.86440 22.81719 23.81149
```

4.3.2 Random-Intercept MLM: Adding Independent Variables

Extending the null model in Eq. (4.5), we can add independent variables to investigate the predictability of the independent variables to the dependent variable while accounting for the nested structure. The three-level null MLM in Eq. (4.5) would become a three-level random-intercept MLM as in Eq. (4.6):

$$y_{ijk} = \beta_0 + V_{00k} + U_{0jk} + \beta_1 x_{ijk} + \epsilon_{ijk}. \tag{4.6}$$

In this model, an extra fixed-effects term $\beta_1 x_{ijk}$ is added to estimate the relationship between the independent variable x (i.e., *RavensTest1*) and the outcome variable y (i.e., *English*). The random-effects term is still the same as it is in the null model as $V_{00k} + U_{0jk} + \epsilon_{ijk}$.

To extend the null model to the random-intercept model, we include the independent variable, *RavensTest1*, to predict the student's *English* test score. The *R* implementation for this is as follows:

```
# fit 3-level HLM with one independent variable
Mod3.1 = lme(fixed = English~RavensTest1, random=~1|School/Class, data = d1)
# print the model fit
summary(Mod3.1)
```

```
## Linear mixed-effects model fit by REML
##   Data: d1
##        AIC      BIC    logLik
##    10333.36 10358.6 -5161.678
##
## Random effects:
##  Formula: ~1 | School
##          (Intercept)
## StdDev:    5.307196
##
##   Formula: ~1 | Class %in% School
##          (Intercept) Residual
## StdDev:    5.881659 20.41193
##
## Fixed effects: English ~ RavensTest1
##                   Value Std.Error   DF t-value p-value
## (Intercept) -0.5467929 2.9956363 1060 -0.18253  0.8552
## RavensTest1  1.9143309 0.1097226 1060 17.44701  0.0000
```

```
##  Correlation:
##              (Intr)
## RavensTest1 -0.918
##
## Standardized Within-Group Residuals:
##         Min          Q1         Med          Q3         Max
## -2.7644107 -0.7381326   0.0251470   0.7801418   2.3514947
##
## Number of Observations: 1154
## Number of Groups:
##             School Class %in% School
##                 49                 93
```

Based on the summary of the model fitting, we can determine:

- Best model fit:

The AIC and BIC values dropped from 10598.31, 10618.51 in the null model to 10333.36, 10358.6 in the random-intercept model after *RavensTest1* is included in the model. As lower values indicate better fit, the random-intercept model is a better fit for the data in comparison to the null model.

- Fixed-effects parameter estimates:

In the random-intercept model with *RavensTest1* as an independent variable, the slope is estimated as $\hat{\beta}_1 = 1.914$, with SE $= 0.109$, t-value $= 17.447$, and p-value < 0.0001, which is highly statistically significant. This means that the *RavensTest1* is a significant predictor of student's *English* scores with higher the *RavensTest1* score, the higher their *English* scores.

- Random-effects parameter estimates:

The same discussion can be made to the estimation of random effects. We leave this to the readers to extract their estimated values. Using the *intervals* command, we can obtain confidence intervals for both the fixed and random effects in the model as follows:

```
# Extract the 95% confidence intervals
intervals(Mod3.1)
```

```
## Approximate 95% confidence intervals
##
##  Fixed effects:
##                   lower        est.      upper
## (Intercept)  -6.424844  -0.5467929  5.331258
## RavensTest1   1.699033   1.9143309  2.129629
## attr(,"label")
## [1] "Fixed effects:"
##
##  Random Effects:
```

```
##    Level: School
##                        lower      est.     upper
## sd((Intercept))  3.020829 5.307196 9.324038
##    Level: Class
##                        lower      est.     upper
## sd((Intercept))  3.676633 5.881659 9.409127
##
##  Within-group standard error:
##    lower      est.     upper
## 19.56139 20.41193 21.29945
```

4.3.3 Random-Slope MLM

To further extend the random-intercept model, we can add the random-slope to fit the general MLM model in model (4.4). In this general model, we allow the slopes (i.e., impact) between *RavensTest1* and *English* to vary across the level 3 *School* and level 2 *Class* nested within *School*.

Due to the additional parameters from random-slope MLM, the default *lme* setting on optimization *nlminb* (i.e., nonlinear unconstrained and box-constrained optimizations using PORT routines) in *opt=c("nlminb")* did not converge. We therefore need to modify the default convergence criteria using *control* to change the default setting of *opt=c("nlminb")* to *opt= c("optim")*. The *optim* is a general-purpose optimization based on Nelder–Mead, quasi-Newton, and conjugate-gradient algorithms, which can produce better modeling fitting.

This can be implemented as follows:

```
# Fit random-slope MLM
Mod3.2 = lme(fixed = English~RavensTest1,
    random = ~RavensTest1|School/Class, data = d1,
    control= lmeControl(opt = c("optim")))
# Print the summary of model fitting
summary(Mod3.2)
```

```
## Linear mixed-effects model fit by REML
##  Data: d1
##        AIC      BIC    logLik
##   10340.65 10386.09 -5161.325
##
## Random effects:
##  Formula: ~RavensTest1 | School
##  Structure: General positive-definite, Log-Cholesky parametrization
##            StdDev     Corr
## (Intercept) 10.6986241 (Intr)
```

```
## RavensTest1   0.3382625 -0.866
##
##   Formula: ~RavensTest1 | Class %in% School
##   Structure: General positive-definite, Log-Cholesky parametrization
##                 StdDev        Corr
## (Intercept)    5.76191425  (Intr)
## RavensTest1    0.02901075  -0.108
## Residual      20.33681552
##
## Fixed effects: English ~ RavensTest1
##                 Value Std.Error   DF   t-value p-value
## (Intercept) -1.216055  3.323965 1060 -0.365845  0.7146
## RavensTest1  1.940309  0.121838 1060 15.925318  0.0000
##   Correlation:
##               (Intr)
## RavensTest1 -0.934
##
## Standardized Within-Group Residuals:
##          Min          Q1         Med          Q3         Max
## -2.74048566 -0.73887068  0.01511409  0.78059478  2.35298741
##
## Number of Observations: 1154
## Number of Groups:
##               School Class %in% School
##                   49                93
```

From this random-slope model, we can determine:

- Fixed-effects parameter estimates:

In this random-slope model, the slope is estimated as $\hat{\beta}_1 = 1.940$, with SE $= 0.122$, t-value $= 15.925$, and p-value < 0.0001, which is still highly statistically significant.

- Model fit:

The AIC and BIC values increased slightly from 10333.36, 10358.6 in the random-intercept model to 10340.65 10386.09 in this random-slope model after the slopes on *RavensTest1* to *English* are assumed to be random. This indicates that the added random-slope did not improve the model fit. A formal statistical Chi-Square test will be illustrated in the next section on model selection.

- Random-effects parameter estimates:

As seen from the summary, the estimated level 2 between-cluster slope standard deviation is $\hat{\tau}_{1.2} = 0.029$ and the level 3 between-cluster slope standard deviation is $\tau_{1.3} = 0.338$. Both are very small indicating that the variations in slopes at both levels might be negligible. In this case, the random-intercept model is a better choice than the random-slope model to fit the data.

- Model selection:

In Sect. 4.3.5, we will refit these models with maximum likelihood estimation and use likelihood ratio tests to test these models. We will see that the random-slope model is in fact not statistically significantly different from the random-intercept model. This confirms the model summary in this section.

4.3.4 Moderation Analysis with Three-Level MLM

The relationship between the scores of *RavensTest1* and the scores of *English* might be moderated by *Gender*. To explore the gender differences in the scores of *English* while accounting for the impact of the score of *RavensTest1*, we can hypothesize that gender's impact on *English* may differ across schools and across classrooms. As such, we must propose a model in which the gender coefficient can vary across level 3 *School* and level 2 *Class* in a three-level MLM.

Since we know from the previous analysis that the random-slope model is as good as the random-intercept model to fit this data, we will add the interaction using the random-intercept model. This can be implemented as follows:

```
# Fit the moderation model with interaction
Mod3.3 = lme(fixed = English~RavensTest1*Gender,
        random = ~1|School/Class, data = d1)
# Print the summary of model fitting
summary(Mod3.3)
```

```
## Linear mixed-effects model fit by REML
##  Data: d1
##        AIC      BIC    logLik
##    10262.56 10297.9 -5124.281
##
## Random effects:
##  Formula: ~1 | School
##         (Intercept)
## StdDev:    4.770976
##
##  Formula: ~1 | Class %in% School
##         (Intercept) Residual
## StdDev:    5.788183 19.80071
##
## Fixed effects: English ~ RavensTest1 * Gender
##                        Value Std.Error   DF   t-value p-value
## (Intercept)         3.428607  3.748275 1058  0.914716  0.3605
## RavensTest1         1.959837  0.141630 1058 13.837724  0.0000
## Gender            -10.091595  5.315539 1058 -1.898508  0.0579
## RavensTest1:Gender -0.018865  0.206276 1058 -0.091453  0.9271
##  Correlation:
##                    (Intr) RvnsT1 Gender
## RavensTest1        -0.941
## Gender             -0.636  0.636
```

```
## RavensTest1:Gender   0.623 -0.660 -0.974
##
## Standardized Within-Group Residuals:
##          Min           Q1          Med           Q3          Max
## -2.63841154 -0.71356520   0.01966517   0.73961589   2.54378700
##
## Number of Observations: 1154
## Number of Groups:
##            School Class %in% School
##            49                93
```

As seen from the model fit, the interaction term is not statistically significant,
i.e., in *RavensTest1:Gender* the estimate $= -0.018865$ with a *p*-value $= 0.9271$.
We reduce the interaction model to the main-effects model as follows:

```
# Reduced model without interaction
Mod3.4 = lme(fixed = English~RavensTest1+Gender,
        random = ~1|School/Class, data = d1)
# Print the summary of model fitting
summary(Mod3.4)
```

```
## Linear mixed-effects model fit by REML
##  Data: d1
##        AIC       BIC    logLik
##   10259.25 10289.54 -5123.626
##
## Random effects:
##  Formula: ~1 | School
##         (Intercept)
## StdDev:    4.761914
##
##  Formula: ~1 | Class %in% School
##         (Intercept) Residual
## StdDev:    5.793704 19.79178
##
## Fixed effects: English ~ RavensTest1 + Gender
##                   Value Std.Error   DF   t-value p-value
## (Intercept)    3.643237 2.9301883 1059  1.243346   0.214
## RavensTest1    1.951263 0.1063897 1059 18.340711   0.000
## Gender       -10.564965 1.2093510 1059 -8.736061   0.000
##  Correlation:
##             (Intr) RvnsT1
## RavensTest1 -0.902
## Gender      -0.163 -0.040
##
## Standardized Within-Group Residuals:
##          Min           Q1          Med           Q3          Max
## -2.63578421 -0.71488725   0.02097198   0.73627144   2.54630700
##
## Number of Observations: 1154
## Number of Groups:
##            School Class %in% School
##            49                93
```

Table 4.1 Summary of model fitting

	Value	Std.Error	DF	t-value	p-value
(Intercept)	3.643	2.930	1059	1.243	0.214
RavensTest1	1.951	0.106	1059	18.341	0.000
Gender	-10.565	1.209	1059	-8.736	0.000

As seen from the summary,

- Estimated random effects:

The within-cluster standard deviation is estimated at $\hat{\sigma} = 19.792$. The estimated level 2 between-cluster intercept standard deviation is $\hat{\tau}_{0.2} = 5.794$ and the level 3 between-cluster intercept standard deviation is $\tau_{0.3} = 4.762$.

- Estimated fixed effects:

The estimated fixed effects are reported in *English ~ RavensTest1 + Gender* as summarized in Table 4.1 as follows:

```
knitr::kable(
round(summary(Mod3.4)$tTable,3), caption = 'Summary of Model Fitting',
    booktabs = TRUE
)
```

which shows that the overall intercept is estimated $\hat{\beta}_0 = 3.643$ and not statistically significant (i.e., $p - \text{value} = 0.214$). The estimated slope is $\hat{\beta}_1 = 1.951$, which is highly statistically significant (i.e., $p - \text{value} = 1.838e - 65$), and the estimated *Gender* effect is $\hat{\beta}_2 = -10.565$, which is also highly statistically significant (i.e., $p - \text{value} = 9.282e - 18$). Since *boy* is coded as 1 in *Gender*, this means that the overall score of the *English* test for *boy*s is 10.565 points lower than that for *girl*s, accounting for the *RavensTest1*.

This can be graphically illustrated with *xyplot*, which adds a regression line to Fig. 4.2 as follows:

```
# Add panel to xyplot
xyplot( English~RavensTest1|as.character(Gender),
        group=School, data=d1,type=c("p","r"),
        ylim=c(-10, 90), xlab="Ravens Test Score", ylab="English Score",
        panel = function(x,y,...){
        panel.xyplot(x,y,...)
        panel.abline(lm(y~x),lwd=3)
        panel.abline(h=coef(lm(y~x))[1],lwd=2, lty=4,col="blue")
    })
```

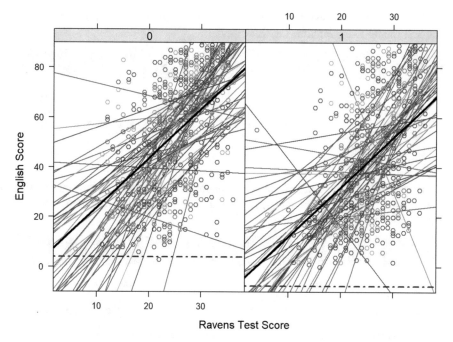

Fig. 4.2 The Raven's test score and the English score by gender

4.3.5   *Model Selection*

For model selection, we can refit the above models with *method="ML"* (i.e., *maximum likelihood estimation*), like what we did in two-level MLM, without printing all the model summaries:

```
# refit the null model using ML
Mod3.0.2 =update(Mod3.0, method="ML")
#summary(Mod3.0.2)
# Refit the Random-Intercept MLM
Mod3.1.2 =update(Mod3.1, method="ML")
#summary(Mod3.1.2)
# Refit the Random-Slope MLM
Mod3.2.2 =update(Mod3.2, method="ML")
#summary(Mod3.2.2)
# Refit the random-intercept MLM with Gender main effect
Mod3.4.2 = update(Mod3.4, method="ML")
#summary(Mod3.2)
```

We can then use these fitted models for model selection. For example, to compare the null model to the random-intercept model:

```
# model comparison between null model v.s. random-intercept model
anova(Mod3.0.2, Mod3.1.2)

##            Model df      AIC      BIC    logLik    Test L.Ratio
## Mod3.0.2      1  4 10600.78 10620.98 -5296.389
## Mod3.1.2      2  5 10332.94 10358.20 -5161.473 1 vs 2 269.833
##            p-value
## Mod3.0.2
## Mod3.1.2  <.0001
```

We see that that the random-intercept model fits better than the null model with a *p*-value < 0.0001. Consequently, we can compare the random-intercept model to the random-slope model:

```
# model comparison  between random-intercept and random-slope models
anova(Mod3.1.2, Mod3.2.2)

##            Model df     AIC      BIC   logLik Test    L.Ratio p-value
## Mod3.1.2      1  5 10332.94 10358.20 -5161.473
## Mod3.2.2      2  9 10340.42 10385.88 -5161.210 1vs2 0.5261411 0.9709
```

We can see that these two models are virtually identical with *p*-value $= 0.9709$, which is why we use the random-intercept model for moderation analysis to investigate the *Gender* effect. To test the *Gender* effect, we can also use the model comparison technique as follows:

```
# model comparison between random-intercept and moderation model
anova(Mod3.1.2, Mod3.4.2)

##            Model df     AIC      BIC   logLik Test L.Ratio p-value
## Mod3.1.2      1  5 10332.94 10358.20 -5161.473
## Mod3.4.2      2  6 10260.88 10291.18 -5124.439 1vs2 74.06667 <.0001
```

This indicates that the moderation model *Mod3.4.2* is better than the random-intercept model in *Mod3.1.2*, which further explains that *Gender* is a significant predictor between the *English* score and *RavensTest1*.

4.4 Exercises

1. Using the same dataset in this chapter,

   - analyze the *Math* score following the steps in this chapter,
   - analyze the data from third junior year (i.e., subset the data with *SchoolYear==2*) following the steps in the chapter.

2. Using the *Health Behavior in School-Aged Children (HBSC)* dataset in Appendix A.3,

- perform a three-level MLM with *Student* (named as *CASEID*) as the level 1, *School* (named as *SCHL_ID*) as level 2, and *District* (named as *DIST_ID*) as level 3 to test whether *Physical Activity* (named as *physact*) is significantly related to *Body Mass Index* (named as *BMI*),
- test whether this relationship between *Physical Activity* (named as *physact*) and *Body Mass Index* (named as *BMI*) is different by *Student's sex* (named as *sex*) while accounting for the three-level nested structure.

Chapter 5
Longitudinal Data Analysis

Longitudinal data is in fact a special type of multi-level data where repeated longitudinal data points can be observed for each individual. In this situation, the repeated longitudinal data points can be viewed as nested within each individual, and subsequently, the individuals can be nested within classrooms, and classrooms can be further nested within schools, etc. With the nested structure, longitudinal data is just to add another level to the existing multi-level data structure.

Therefore, we can use multi-level modeling to analyze longitudinal data. In this chapter, we will present the analysis for longitudinal data from Appendix A on *Longitudinal Breast Cancer Post-surgery Assessment* in Sect. A.5. We will *reshape* the *wide* (i.e., person-level) data format to the *long* (i.e., longitudinal person-period) data format, so we can perform longitudinal data analysis with random-intercept and random-slope models (i.e., mixed-effects models). The purpose of the original analysis was to identify change over the 8-month period in individual perceptions of *mood* and *social adjustment* by women who recently underwent breast cancer surgery.

In this chapter, we will only analyze the outcome data on *social adjustment* as demonstration of longitudinal data analysis and leave the other outcome, *mood*, as a practice exercise. We will further use this dataset to demonstrate the addition of *age* and *type* of surgical treatment as possible predictors that may account for any change in the individual growth trajectories (i.e., intercept and slope) of *social adjustment*.

Again, we will use *R* package *nlme* (Pinheiro and Bates 2000) for the model fitting in this chapter.

5.1 Data on Breast Cancer Post-surgery Assessment

Let us load the data into our *R* session:

© The Author(s), under exclusive license to Springer Nature Switzerland AG 2021
D.-G. (Din) Chen, J. K. Chen, *Statistical Regression Modeling with R*,
Emerging Topics in Statistics and Biostatistics,
https://doi.org/10.1007/978-3-030-67583-7_5

```
# Read the data into R
dHK = read.csv("hkcancer.csv", header=T)
# Check the data dimensions
dim(dHK)
```

```
## [1] 405   10
```

```
# Check the variables
names(dHK)
```

```
## [1] "ID"       "Mood1"   "Mood4"   "Mood8"   "SocAdj1"
## [6] "SocAdj4" "SocAdj8" "Age"     "Age2"    "SurgTx"
```

```
# Extract variables associated with *Social Adjustment*
dHK.Soc = dHK[, c("ID","SocAdj1","SocAdj4","SocAdj8","Age2","SurgTx")]
# Check the data dimensions
dim(dHK.Soc)
```

```
## [1] 405   6
```

```
# Summary of the data
summary(dHK.Soc)
```

```
##       ID            SocAdj1           SocAdj4
##  Min.   :  1   Min.   : 33.00   Min.   : 51.86
##  1st Qu.:102   1st Qu.: 96.74   1st Qu.: 95.91
##  Median :203   Median :101.06   Median :100.03
##  Mean   :203   Mean   :100.91   Mean   :100.42
##  3rd Qu.:304   3rd Qu.:106.22   3rd Qu.:105.39
##  Max.   :405   Max.   :132.00   Max.   :145.20
##                NA's   :33       NA's   :69
##     SocAdj8          Age2            SurgTx
##  Min.   : 64.82   Min.   :0.0000   Min.   :0.0000
##  1st Qu.: 96.56   1st Qu.:0.0000   1st Qu.:1.0000
##  Median :100.22   Median :0.0000   Median :1.0000
##  Mean   :100.60   Mean   :0.4938   Mean   :0.8049
##  3rd Qu.:105.35   3rd Qu.:1.0000   3rd Qu.:1.0000
##  Max.   :133.14   Max.   :1.0000   Max.   :1.0000
##  NA's   :60
```

```
# Print the first 6 observations to show "wide" data form
head(dHK.Soc)
```

```
##   ID   SocAdj1    SocAdj4   SocAdj8 Age2 SurgTx
## 1  1  95.90625        NA        NA    1      1
## 2  2 114.88890 105.11110  90.44444    0      1
```

```
## 3   3    80.66667    95.33333    95.33333      0        1
## 4   4  112.83870   108.58060    99.00000      1        1
## 5   5  115.00000   105.00000   101.00000      0        1
## 6   6  106.45160   114.96770   107.51610      0        0
```

In this data *dHK*, we have the data stored in the *wide* format. In the *wide* format, each row represents an individual woman and each column represents a recorded measurement/variable for that woman at different time points.

To analyze this data with longitudinal modeling, we first need to restructure the person-level data in *wide* format into person-period format (i.e., *long* format or longitudinal format). In the *long* format, more rows are added for each individual woman to account for each *social adjustment* measurement at the three time points. Therefore, the number of rows that correspond to an individual woman will reflect the total number of measurements (i.e., 3 in this case) made for the *social adjustment*. In this dataset, the *social adjustment* was measured three times at 1, 4, and 8 months post-surgery. Therefore, each woman will have three rows that correspond to this variable.

To convert data from a *wide* format (i.e., person-level) to a *long* format (i.e., person-period), three steps are needed:

1. all time-variant variables (e.g., all *SocAdj* measurements over the three time points) must be stacked over the three time periods,
2. all time-invariant variables must be copied into the new data file multiple times based on the number of time points (i.e., three times),
3. an extra *time* variable must be created to identify the time.

In *R*, the function *reshape* converts data from *wide* format to *long* format and vice versa. To convert the *dHK.Soc* from a *wide* format to a *long* format, we can make use of the following *R* chunk format:

```
# Call *reshape* to convert the data format
dSoc = reshape(dHK.Soc, direction = "long", idvar = "ID", timevar = "Time",
               varying = c("SocAdj1","SocAdj4","SocAdj8"), sep="")
# Check the dimension: should be 3 times
dim(dSoc)
```

```
## [1] 1215     5
```

```
# Print the first 6 observations
head(dSoc)
```

```
##      ID Age2 SurgTx Time     SocAdj
## 1.1   1    1      1    1   95.90625
## 2.1   2    0      1    1  114.88890
## 3.1   3    0      1    1   80.66667
## 4.1   4    1      1    1  112.83870
## 5.1   5    0      1    1  115.00000
## 6.1   6    0      0    1  106.45160
```

```
# Remove the missing values for longitudinal modeling
dSoc = dSoc[complete.cases(dSoc),]
# Check the dimension again
dim(dSoc)
```

```
## [1] 1053     5
```

```
# Print the data summary
summary(dSoc)
```

```
##       ID              Age2              SurgTx
##  Min.   :  1.0    Min.   :0.0000    Min.   :0.000
##  1st Qu.:101.0    1st Qu.:0.0000    1st Qu.:1.000
##  Median :205.0    Median :0.0000    Median :1.000
##  Mean   :203.8    Mean   :0.4729    Mean   :0.793
##  3rd Qu.:307.0    3rd Qu.:1.0000    3rd Qu.:1.000
##  Max.   :405.0    Max.   :1.0000    Max.   :1.000
##       Time             SocAdj
##  Min.   :1.000    Min.   : 33.00
##  1st Qu.:1.000    1st Qu.: 96.46
##  Median :4.000    Median :100.22
##  Mean   :4.251    Mean   :100.65
##  3rd Qu.:8.000    3rd Qu.:105.60
##  Max.   :8.000    Max.   :145.20
```

As exercise, we can use *reshape* to convert the *long* format data into the *wide* format data to check the original *wide* format of data in *dHK.Soc* as follows:

```
# convert back from "long" format to "wide" format
datback = reshape(dSoc, direction="wide")
# Check the wide format data
head(datback)
```

```
##        ID Age2 SurgTx    SocAdj1    SocAdj4    SocAdj8
## 1.1   1    1      1   95.90625         NA         NA
## 2.1   2    0      1  114.88890  105.11110   90.44444
## 3.1   3    0      1   80.66667   95.33333   95.33333
## 4.1   4    1      1  112.83870  108.58060   99.00000
## 5.1   5    0      1  115.00000  105.00000  101.00000
## 6.1   6    0      0  106.45160  114.96770  107.51610
```

You can examine the new dataframe *datback*, and it is in the same format as *dHK.Soc*.

5.2 Graphic Analysis

Once the data is converted to long format, it can be plotted and analyzed. We start the data analysis with data graphics to illustrate the data and reveal the data trends longitudinally.

As a preliminary data analysis, we plot the *SocAdj* as a function of *Time* for each woman as seen in Fig. 5.1 with the following *R* code chunk:

```
# load the R package
library(lattice)
# Call *xyplot* to plot the longitudinal data
xyplot(SocAdj~Time|as.character(SurgTx),
       ylab="Social Adjustment",group=ID,type=c("p","l"), data=dSoc)
```

In this *R* code chunk, we use the *R* package *lattice* and call the function *xyplot* to plot the *SocAdj* to *Time* for each woman (i.e., *group=ID*) by *SurgTx*. The *p* and *l* in *type=c("p","l")* are used to plot the data with *points* and linked with *lines*.

This is also known as a *spaghetti* plot, with each person representing one line of data. One purpose of analyzing the data visually is to determine the nature of the relationship between data points over time (e.g., linear and loglinear). From this figure, there appears to be a linear relationship in the data, with only a few instances of a nonlinear relationship.

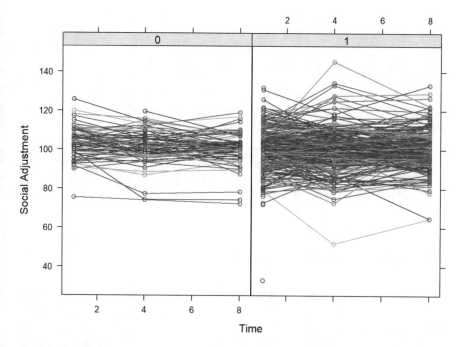

Fig. 5.1 Social adjustment as a function of time for each group of women

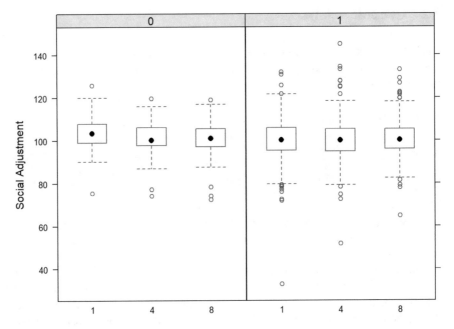

Fig. 5.2 Distribution of social adjustment as a function of time for all women

Additionally, we can see from this figure that there are more women in the *mastectomy* surgical treatment group than those in the *lumpectomy* group (note that 0 = *lumpectomy* and 1 = *mastectomy*). It seems that there are no obvious patterns in *SocAdj*, but we will find out more in our statistical analysis.

Further trends in the data may be represented by generating boxplots, as seen in Fig. 5.2. From this figure, the data at each time point appears to be normally distributed with few outliers. This data exploration further justifies the longitudinal data analysis with a normally distributed error structure.

```
# Call *bwplot* to plot the longitudinal data
bwplot(SocAdj~as.factor(Time)|as.character(SurgTx),
       ylab="Social Adjustment", data=dSoc)
```

5.3 Response Feature Analysis

From Fig. 5.1, a linear growth model for each woman seems reasonable. That is, for each woman i (keep in mind that the individual woman i becomes the level 2 in longitudinal data), her *SocAdj* observed at *Time* (i.e., t, repeated measurements for each woman i are at the level 1 in longitudinal data) can be reasonably modeled as a linear regression model:

$$y_{it} = \beta_{i0} + \beta_{i1}Time_{it} + \epsilon_{it}, \tag{5.1}$$

where y_{it} denotes the outcome of *SocAdj* at time t for woman i, β_{i0} and β_{i1} represent the parameters of intercept and slope of the longitudinal linear model for the ith woman, and ϵ_{it} is the error term that is assumed to be normally distributed with a mean of 0 and within-woman standard deviation σ.

To extract the response features, we estimate the *slope*(i.e. β_{i1}) and *intercept* (i.e., β_{i0}) for each woman as depicted in Fig. 5.1. To do so, we can loop over the 405 women to fit a linear regression and then extract the intercepts and slopes to make a dataframe as follows:

```
# Extract the IDs for all women
Woman.IDs = unique(dSoc$ID)
# Number of women
num.Subj = length(Woman.IDs)
# initiate the intercept and slope
beta0 = beta1 = trt= numeric(num.Subj)

# loop-over
for(i in 1:num.Subj){
# fit regression model for all Women
mod = lm(SocAdj~Time, dSoc[dSoc$ID==Woman.IDs[i],])
# extract the intercept and slope
beta0[i] = coef(mod)[1]; beta1[i] = coef(mod)[2]
# extract their treatment (denoted by "trt")
trt[i]   = dSoc[dSoc$ID==Woman.IDs[i],"SurgTx"][1]
}

# make a dataframe "dat.coef"
dat.coef=data.frame(ID=Woman.IDs,SurgTx= trt, beta0=beta0, beta1=beta1)
# Print the data for the first 6 women to see the beta0 and beta1
head(dat.coef)
```

```
##   ID SurgTx     beta0      beta1
## 1  1      1  95.90625         NA
## 2  2      1 118.65466 -3.50150351
## 3  3      1  81.85586  1.98198108
## 4  4      1 115.47120 -1.99956216
## 5  5      1 115.43243 -1.94594595
## 6  6      0 109.45813  0.04315405
```

```
# print the summary
summary(dat.coef)
```

```
##       ID             SurgTx          beta0           beta1
## Min.   :  1.00   Min.   :0.0000   Min.   : 33.00   Min.   :-3.9049
## 1st Qu.: 99.25   1st Qu.:1.0000   1st Qu.: 95.77   1st Qu.:-0.8919
## Median :204.50   Median :1.0000   Median :100.68   Median :-0.1429
## Mean   :202.89   Mean   :0.7953   Mean   :100.96   Mean   :-0.1500
## 3rd Qu.:304.75   3rd Qu.:1.0000   3rd Qu.:107.06   3rd Qu.: 0.6541
```

```
##   Max.    :405.00   Max.    :1.0000   Max.    :132.00   Max.    : 6.0054
##                                                          NA's    :33
```

We can calculate some statistics for testing intercepts and slopes, as listed below:

```
# Mean for intercept2
mean(dat.coef[,3])
```

```
## [1] 100.9606
```

```
# Mean for the slopes
mean(dat.coef[,4], na.rm=T);
```

```
## [1] -0.1500093
```

```
# Covariance matrix
va = var(dat.coef[,3:4], na.rm = T);va
```

```
##              beta0      beta1
## beta0 93.904585  -6.949381
## beta1 -6.949381   1.548357
```

```
# Standard deviations
sqrt(va)
```

```
## Warning in sqrt(va): NaNs produced
```

```
##              beta0     beta1
## beta0 9.690438       NaN
## beta1      NaN  1.24433
```

```
# Correlations
cov2cor(va)
```

```
##              beta0       beta1
## beta0  1.0000000  -0.5763245
## beta1 -0.5763245   1.0000000
```

Furthermore, from this dataframe *dat.coef*, we note that the *beta0* for *SocAdj* varies by approximately 100 units with a mean slope of about −0.15. A bivariate plot of the *beta0* and *beta1* from these women appears in Fig. 5.3. This figure clearly shows that the intercepts for *SocAdj* from these 405 women are sufficiently normally distributed with a mean of 100 and a standard deviation of 9.69 (the histogram at the top), and the slopes are normally distributed with a mean of about −0.15 and a standard deviation of 1.244 (the histogram on the right side). In addition, all the intercepts and slopes are negatively correlated with a correlation coefficient of −0.576. The variations embedded in the intercepts and slopes illustrate the modeling of random effects for both intercept and slope.

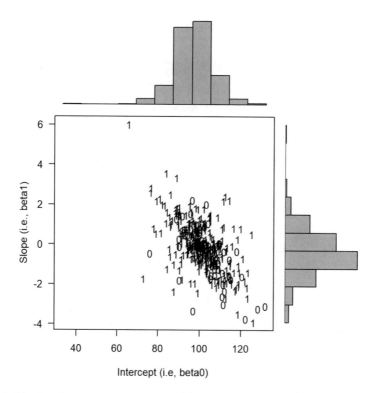

Fig. 5.3 Bivariate distribution of intercepts and slopes from 405 women

```
# Make histogram for both intercept and slope
beta0.hist = hist(beta0,plot=F); beta1.hist = hist(beta1,plot=F)

# make layout for plotting
top = max(c(beta0.hist$counts, beta1.hist$counts))
nf  = layout(matrix(c(2,0,1,3),2,2,byrow=T), c(3,1), c(1,3),T)
par(mar=c(5,4,1,1))
# plot the intercept and slope
plot(beta1~beta0,las=1,dat.coef,xlab="Intercept (i.e., beta0)",
  ylab="Slope (i.e., beta1)",pch=as.character(SurgTx))

par(mar=c(0,4,1,1))
# add the intercept histogram
barplot(beta0.hist$counts, axes=FALSE, ylim=c(0, top), space=0)
par(mar=c(5,0,1,1))
# add the slope histogram
barplot(beta1.hist$counts, axes=FALSE,
        xlim=c(0, top), space=0, horiz=TRUE)
```

We will now model the slope and intercept relationship by a linear regression. In doing so, we fit three models: the first with an interaction term including *SurgTx*,

the second without an interaction term that includes *SurgTx*, and the third without
the *SurgTx* term.

```
# fit model 1 with interaction
mod1.coef = lm(beta1~beta0*SurgTx, dat.coef)
summary(mod1.coef)
```

```
##
## Call:
## lm(formula = beta1 ~ beta0 * SurgTx, data = dat.coef)
##
## Residuals:
##     Min      1Q  Median      3Q     Max
## -3.6540 -0.5495 -0.0278  0.5892  3.5852
##
## Coefficients:
##                Estimate Std. Error t value Pr(>|t|)
## (Intercept)     8.05010    1.37172   5.869 1.02e-08 ***
## beta0          -0.08218    0.01315  -6.249 1.21e-09 ***
## SurgTx         -0.98258    1.51059  -0.650    0.516
## beta0:SurgTx    0.01138    0.01456   0.781    0.435
## ---
## Signif. codes:  0 '***' 0.001 '**' 0.01 '*' 0.05 '.' 0.1 ' ' 1
##
## Residual standard error: 1.017 on 349 degrees of freedom
##   (33 observations deleted due to missingness)
## Multiple R-squared:  0.3372, Adjusted R-squared:  0.3315
## F-statistic:  59.2 on 3 and 349 DF,  p-value: < 2.2e-16
```

```
# fit model 2 without interaction
mod2.coef = lm(beta1~beta0+SurgTx, dat.coef)
summary(mod2.coef)
```

```
##
## Call:
## lm(formula = beta1 ~ beta0 + SurgTx, data = dat.coef)
##
## Residuals:
##     Min      1Q  Median      3Q     Max
## -3.7126 -0.5506 -0.0079  0.6350  3.5117
##
## Coefficients:
##               Estimate Std. Error t value Pr(>|t|)
## (Intercept)   7.085667   0.598442   11.84   <2e-16 ***
## beta0        -0.072896   0.005645  -12.91   <2e-16 ***
## SurgTx        0.193213   0.134208    1.44    0.151
## ---
## Signif. codes:  0 '***' 0.001 '**' 0.01 '*' 0.05 '.' 0.1 ' ' 1
##
## Residual standard error: 1.017 on 350 degrees of freedom
##   (33 observations deleted due to missingness)
## Multiple R-squared:  0.3361, Adjusted R-squared:  0.3323
## F-statistic: 88.59 on 2 and 350 DF,  p-value: < 2.2e-16
```

```
# fit model 3 without SurgTx
mod3.coef = lm(beta1~beta0, dat.coef)
summary(mod3.coef)
```

```
##
## Call:
## lm(formula = beta1 ~ beta0, data = dat.coef)
##
## Residuals:
##      Min      1Q  Median      3Q     Max
## -3.7039 -0.5506 -0.0275  0.6178  3.5126
##
## Coefficients:
##               Estimate Std. Error t value Pr(>|t|)
## (Intercept)   7.350724   0.570286   12.89   <2e-16 ***
## beta0        -0.074005   0.005601  -13.21   <2e-16 ***
## ---
## Signif. codes:  0 '***' 0.001 '**' 0.01 '*' 0.05 '.' 0.1 ' ' 1
##
## Residual standard error: 1.018 on 351 degrees of freedom
##    (33 observations deleted due to missingness)
## Multiple R-squared:  0.3321, Adjusted R-squared:  0.3302
## F-statistic: 174.6 on 1 and 351 DF,  p-value: < 2.2e-16
```

From model 1, the interaction (i.e., Intercept:TRTB) is not statistically significant. Therefore, we reduce to the main effect model in model 2. In model 2, the main effect of *SurgTx* is also not significant, so we further reduce to model 3 without *SurgTx*. In model 3, the relationship between the *beta1*s and the *beta0*s is significant indicating that there is a significant correlation between the *beta1* and the *beta0*. This result confirms the graphical illustration in Fig. 5.3.

Further analysis of the difference between "*SurgTx*" treatment (0 = *lumpectomy*; 1 = *mastectomy*) can be performed using the *t-test* as

```
# test slope difference
t.test(beta1~SurgTx, dat.coef)
```

```
##
##   Welch Two Sample t-test
##
## data:  beta1 by SurgTx
## t = -2.668, df = 115.03, p-value = 0.008733
## alternative hypothesis: true difference in means is not equal to 0
## 95 percent confidence interval:
##   -0.7484905 -0.1106398
## sample estimates:
## mean in group 0 mean in group 1
##      -0.48952401      -0.05995884
```

```
# test intercept difference
t.test(beta0~SurgTx, dat.coef)
```

```
##
##  Welch Two Sample t-test
##
## data:  beta0 by SurgTx
## t = 3.174, df = 144.66, p-value = 0.001837
## alternative hypothesis: true difference in means is not equal to 0
## 95 percent confidence interval:
##  1.394661 5.998307
## sample estimates:
## mean in group 0 mean in group 1
##        103.9006        100.2041
```

In longitudinal data analysis, the response feature analysis extracts fundamental
features from each woman for simple and preliminary data exploration and summarization. This feature analysis provides basic summary information from the data for
simple conclusions in addition to providing directions for further analysis. However,
this analysis also loses information since other features are excluded. A more
efficient analysis is to use all the information from all the data in a comprehensive
manner to capture the longitudinal or repeated measures nature of the data.

5.4 Longitudinal Models Using Function *lme* from *nlme* Package

5.4.1 Random-Slope Model

For this data, we will begin with the random-slope multi-level modeling because of
the visual exploration in Fig. 5.1 of the data and the associated *response feature
analysis* which suggest that both the intercepts and slopes from all women are
randomly distributed with bivariate normal distribution.

Therefore, both the coefficients of β_{i0} and β_{i1} in Eq. (5.1) are random with a
negative correlation as follows:

$$\beta_{i0} = \beta_0 + \nu_{i0}$$
$$\beta_{i1} = \beta_1 + \nu_{i1}, \tag{5.2}$$

where β_0 and β_1 are the overall fixed effects to describe the overall group effects
from these 405 women, and νs are the level 2 random errors associated with the two
parameters (i.e., random effects) from each woman to the overall (β_0, β_1), which is
assumed to be bivariate normally distributed as

$$\begin{pmatrix} \nu_{i0} \\ \nu_{i1} \end{pmatrix} \sim N\left(\begin{pmatrix} 0 \\ 0 \end{pmatrix}, \begin{pmatrix} \tau_0^2 & \tau_0\tau_1\rho_{01} \\ \tau_0\tau_1\rho_{01} & \tau_1^2 \end{pmatrix}, \right)$$

which is the between-women variance–covariance matrix, where τ_0^2 and τ_1^2 are the
between-women variances and ρ_{01} is the between-women correlation.

5.4.2   Fitting Random-Slope Model Using *lme*

The *response feature analysis*, however, was done for each woman *without* using other women's data information. For a comprehensive analysis, the mixed-effects (i.e., both random-intercept and random-slope) model is designed for this purpose. In this model, we allow both intercept and slope to vary among all women.

Using the restructured data *dSoc*, the following *R* syntax fits the random-slope model for *SocAdj*:

```
# Model 1: Growth Model (Random-intercept and random-slope)
mSoc1 = lme(fixed = SocAdj~Time, random = ~Time|ID, data = dSoc )
# Print the model summary
summary(mSoc1)
```

```
## Linear mixed-effects model fit by REML
##  Data: dSoc
##        AIC      BIC    logLik
##   7392.175 7421.92 -3690.087
##
## Random effects:
##  Formula: ~Time | ID
##  Structure: General positive-definite, Log-Cholesky parametrization
##             StdDev    Corr
## (Intercept) 8.7838278 (Intr)
## Time        0.6154577 -0.505
## Residual    5.4936658
##
## Fixed effects: SocAdj ~ Time
##                 Value Std.Error  DF  t-value p-value
## (Intercept) 100.8892 0.5420049 666 186.1407  0.0000
## Time         -0.0963 0.0678580 666  -1.4192  0.1563
##  Correlation:
##         (Intr)
## Time -0.602
##
## Standardized Within-Group Residuals:
##        Min         Q1        Med         Q3        Max
## -3.6421826 -0.4327027  0.0123785  0.4345529  4.5370145
##
## Number of Observations: 1053
## Number of Groups: 386
```

As seen from the above fitting, the estimated overall intercept is $\hat{\beta}_0 = 100.889$, which is statistically significant. However, the slope $\beta_1 = -0.096$ is not statistically significant (p-value = 0.1563). The estimated within-women standard deviation is $\hat{\sigma} = 5.494$, and the estimated between-women standard deviation for intercept is $\hat{\tau}_0 = 8.784$ and is $\hat{\tau}_1 = 0.615$ for the slope. The estimated correlation coefficient between random-intercept and random-slope is $\hat{r}_{01} = -0.505$.

5.4.3   Fitting Random-Intercept Model Using lme

For model selection, let us fit a random-intercept only model to see whether or not
we can simplify the above model. In this model, we allow only the *intercept* to vary
among all women and fix the *slope* for all women. This means that all women will
have the same time trajectory, which is quite a restrictive assumption.

This can be easily done by changing *random = ~Time|ID* (i.e., both random-slope
and random-intercept) to *random = ~1|ID* (i.e., random-intercept only) as follows:

```
# Model 2: Growth Model (Random-intercept)
mSoc2 = lme(fixed = SocAdj~Time, random = ~1|ID, data = dSoc )
# Print the summary of model fitting
summary(mSoc2)
```

```
## Linear mixed-effects model fit by REML
##  Data: dSoc
##         AIC      BIC    logLik
##    7396.937 7416.767 -3694.469
##
## Random effects:
##  Formula: ~1 | ID
##          (Intercept) Residual
## StdDev:     7.750296 5.892313
##
## Fixed effects: SocAdj ~ Time
##                  Value Std.Error  DF   t-value p-value
## (Intercept) 100.90027 0.5119184 666 197.10226  0.0000
## Time          -0.10227 0.0638942 666  -1.60062  0.1099
##
## Standardized Within-Group Residuals:
##         Min          Q1          Med          Q3         Max
## -4.21460074 -0.47204505 -0.01333464  0.44633108  4.24842331
##
## Number of Observations: 1053
## Number of Groups: 386
```

The model comparison between *mSoc1* and *mSoc2* can be done to call *R* function
anova with updated maximum likelihood estimation as follows:

```
## Model selection
anova(update(mSoc1, method="ML"), update(mSoc2, method="ML"))
```

```
##                                   Model df      AIC      BIC
## update(mSoc1, method = "ML")          1  6 7388.792 7418.549
## update(mSoc2, method = "ML")          2  4 7393.454 7413.292
##                                   logLik   Test L.Ratio p-value
## update(mSoc1, method = "ML")    -3688.396
## update(mSoc2, method = "ML")    -3692.727 1 vs 2 8.662093  0.0132
```

It can be seen from this model comparison that *mSoc1* has the smaller *AIC* and
BIC than *mSoc2*. The likelihood ratio χ^2-test yielded a *p-value=0.0125* indicating
that the random-slope model in *mSoc1* is a statistically significantly better model

than the random-intercept model *mSoc2*. So, we can see that *mSoc1* is a better model to fit the data.

5.4.4 Fitting Random-Slope MLM with Moderators

In longitudinal data analysis, we can add either time-variant or time-invariant predictors to the mixed-effects model. For easy illustration, we will include the two time-invariant covariates of *age2* and *SurgTx* to illustrate their effects on *SocAdj*. For longitudinal data analysis for time-variant covariates, we refer the interested readers to the books by (Diggle et al. 2002; Wilson and Lorenz 2015) and (Wilson et al. 2020).

We will include these two covariates in the random-slope model *mSoc1* since this model better fitted the data. To include these two covariates in the model, *mSoc1*, we start with the most general 3-way interaction model as follows:

```
# Model 3: AGE & SURGERY TYPE
mSoc3all =lme(fixed=SocAdj~Time*Age2*SurgTx,random=~Time|ID,data=dSoc)
# Print the summary of model fitting
summary(mSoc3all)
```

```
## Linear mixed-effects model fit by REML
##  Data: dSoc
##        AIC      BIC    logLik
##    7390.84 7450.262 -3683.42
##
## Random effects:
##  Formula: ~Time | ID
##  Structure: General positive-definite, Log-Cholesky parametrization
##             StdDev    Corr
## (Intercept) 8.6874064 (Intr)
## Time        0.5995111 -0.48
## Residual    5.4898246
##
## Fixed effects: SocAdj ~ Time * Age2 * SurgTx
##                    Value Std.Error  DF  t-value p-value
## (Intercept)    103.30215 1.6393435 663 63.01434  0.0000
## Time            -0.41916 0.2014660 663 -2.08054  0.0379
## Age2             0.91620 2.3815805 382  0.38470  0.7007
## SurgTx          -2.36148 1.8385799 382 -1.28440  0.1998
## Time:Age2       -0.04147 0.2977592 663 -0.13928  0.8893
## Time:SurgTx      0.34059 0.2268799 663  1.50118  0.1338
## Age2:SurgTx     -2.56428 2.6701499 382 -0.96035  0.3375
## Time:Age2:SurgTx 0.18951 0.3341578 663  0.56713  0.5708
##  Correlation:
##                (Intr) Time   Age2   SurgTx Tm:Ag2 Tm:SrT Ag2:ST
## Time           -0.596
## Age2           -0.688  0.410
## SurgTx         -0.892  0.531  0.614
## Time:Age2       0.403 -0.677 -0.596 -0.359
## Time:SurgTx     0.529 -0.888 -0.364 -0.595  0.601
## Age2:SurgTx     0.614 -0.366 -0.892 -0.689  0.532  0.410
```

```
## Time:Age2:SurgTx -0.359   0.603   0.531   0.404  -0.891  -0.679  -0.595
##
## Standardized Within-Group Residuals:
##          Min              Q1            Med            Q3           Max
## -3.608109889 -0.443331178  0.006876285  0.433501475  4.531507105
##
## Number of Observations: 1053
## Number of Groups: 386
```

Since the 3-way interaction *Time:Age2:SurgTx* (*p*-value = 0.5708) and the 2-way interaction *Age2:SurgTx* (*p*-value = 0.3375) both are not significant, let us reduce the model to the 1-way interaction with the *Time*:

```
# Model 3: AGE & SURGERY TYPE
mSoc3=lme(fixed=SocAdj~Time+Age2*Time+SurgTx*Time,random=~Time|ID,data=dSoc)
# Print the summary of model fitting
summary(mSoc3)
```

```
## Linear mixed-effects model fit by REML
##  Data: dSoc
##        AIC       BIC     logLik
##   7390.771 7440.308 -3685.386
##
## Random effects:
##  Formula: ~Time | ID
##  Structure: General positive-definite, Log-Cholesky parametrization
##             StdDev    Corr
## (Intercept) 8.684617 (Intr)
## Time        0.596890 -0.482
## Residual    5.490546
##
## Fixed effects: SocAdj ~ Time + Age2 * Time + SurgTx * Time
##                 Value Std.Error  DF  t-value p-value
## (Intercept) 104.26888 1.2936997 664 80.59744  0.0000
## Time         -0.48934 0.1605785 664 -3.04736  0.0024
## Age2         -1.12367 1.0766979 383 -1.04363  0.2973
## SurgTx       -3.57740 1.3329689 383 -2.68378  0.0076
## Time:Age2     0.10932 0.1350169 664  0.80967  0.4184
## Time:SurgTx   0.42918 0.1664121 664  2.57901  0.0101
##  Correlation:
##             (Intr) Time   Age2   SurgTx Tm:Ag2
## Time        -0.596
## Age2        -0.394  0.232
## SurgTx      -0.819  0.489 -0.001
## Time:Age2    0.230 -0.385 -0.593  0.005
## Time:SurgTx  0.487 -0.817  0.005 -0.596 -0.013
##
## Standardized Within-Group Residuals:
##          Min          Q1          Med          Q3          Max
## -3.623925073 -0.441590676  0.008275821  0.429400332  4.537038760
## Number of Observations: 1053
## Number of Groups: 386
```

Since the interaction term of *Time:Age2* is not statistically significant (i.e., *p-value=0.4184*), we remove this interaction and only keep the main-effect model for *age2*. As also seen above, the interaction of *Time:SurgTx* is significant. Therefore, we will keep the interaction for *SurgTx*. The *R* code is as follows:

```
# Model 4: Only main-effect for *Age2*
mSoc4=lme(fixed=SocAdj~Time+Age2+SurgTx*Time,random=~Time|ID,data=dSoc)
# Print the model fitting
summary(mSoc4)
```

```
## Linear mixed-effects model fit by REML
##  Data: dSoc
##        AIC       BIC    logLik
##    7387.26  7431.852  -3684.63
##
## Random effects:
##  Formula: ~Time | ID
##  Structure: General positive-definite, Log-Cholesky parametrization
##              StdDev     Corr
## (Intercept) 8.6862409  (Intr)
## Time        0.5953862  -0.482
## Residual    5.4905300
##
## Fixed effects: SocAdj ~ Time + Age2 + SurgTx * Time
##                  Value Std.Error  DF  t-value p-value
## (Intercept) 104.02818 1.2592370 665  82.61207  0.0000
## Time         -0.43932 0.1481239 665  -2.96586  0.0031
## Age2         -0.60602 0.8667148 383  -0.69921  0.4848
## SurgTx       -3.58241 1.3331143 383  -2.68725  0.0075
## Time:SurgTx   0.43087 0.1663048 665   2.59085  0.0098
##  Correlation:
##             (Intr)  Time    Age2    SurgTx
## Time        -0.565
## Age2        -0.329   0.005
## SurgTx      -0.843   0.532   0.002
## Time:SurgTx  0.503  -0.891  -0.004  -0.596
##
## Standardized Within-Group Residuals:
##            Min            Q1          Med            Q3         Max
## -3.634463392 -0.434337006  0.009989651  0.431551174 4.539175849
## Number of Observations: 1053
## Number of Groups: 386
```

The main effect of *Age2* is also not significant. Therefore, we further remove the *Age2* from *mSoc4* and refit the model as follows:

```
mSoc5=lme(fixed=SocAdj~Time+SurgTx*Time,random=~Time|ID,data=dSoc)
# Print the summary of model fit
summary(mSoc5)
```

```
## Linear mixed-effects model fit by REML
##  Data: dSoc
##        AIC       BIC    logLik
##    7387.299  7426.944  -3685.65
##
## Random effects:
##  Formula: ~Time | ID
##  Structure: General positive-definite, Log-Cholesky parametrization
```

```
## 				StdDev		Corr
## (Intercept)	8.6914250	(Intr)
## Time			0.5962565	-0.486
## Residual		5.4897019
##
## Fixed effects: SocAdj ~ Time + SurgTx * Time
## 						Value Std.Error  DF  t-value p-value
## (Intercept) 103.73835 1.1894180 665 87.21775  0.0000
## Time          -0.43878 0.1481461 665 -2.96182  0.0032
## SurgTx        -3.58092 1.3336001 384 -2.68516  0.0076
## Time:SurgTx    0.43055 0.1663303 665  2.58853  0.0098
##  Correlation:
## 					(Intr)  Time    SurgTx
## Time			-0.598
## SurgTx		-0.892	0.533
## Time:SurgTx	0.533  -0.891  -0.597
##
## Standardized Within-Group Residuals:
## 			Min 			Q1 			Med 			Q3 			Max
## -3.64987767 -0.43517394  0.00559863  0.42752515  4.54845478
##
## Number of Observations: 1053
## Number of Groups: 386
```

As seen from the above fitting, the estimated overall intercept for the surgical group of *SurgTx=0* (i.e., women with *lumpectomy* surgical treatment) is $\hat{\beta}_0 = 103.738$ (statistically significant with *p-value = 0.000*) and the slope is $\beta_1 = -0.439$ (statistically significant with *p-value = 0.0032*). This means that the social adjustment for this group of women decreases as time progresses in the 8-month long time period. The estimated overall intercept for the group *SurgTx=1* (i.e., women with *mastectomy* surgical treatment) is $\hat{\beta}_0 = 103.738 - 3.581 = 100.157$, indicating that the initial social adjustment for this group started 3.581 points lower. The slope $\hat{\beta}_1 = -0.439 + 0.431 = -0.008$ is small and negligible, indicating that the social adjustment for this group of women practically did not change for the 8-month period.

With this model, the estimated within-women standard deviation is $\hat{\sigma} = 5.490$, and the estimated between-women standard deviation for the intercept is $\hat{\tau}_0 = 8.691$ and is $\hat{\tau}_1 = 0.596$ for the slope. The estimated correlation coefficient between random-intercept and random-slope is $\hat{r}_{01} = -0.486$. Thus, the model with *SurgTx* in *mSoc5* fits a little better than the model without *SurgTx* in *mSoc1*.

This can be graphically illustrated with an *R panel* plot that uses *xyplot* to add a regression line in Fig. 5.1, which can be seen in Fig. 5.4:

```
# Add panel to xyplot
xyplot(SocAdj~Time|as.character(SurgTx),group=ID, data=dSoc,type=c("p","l"),
       xlab="Time", ylab="Social Adjustment", ylim=c(70,140),
       panel = function(x,y,...){
       panel.xyplot(x,y,...); panel.abline(lm(y~x),lwd=4, col="black")
    })
```

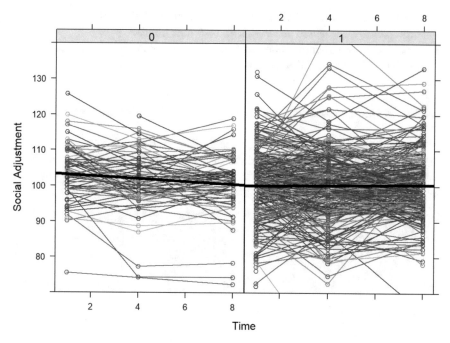

Fig. 5.4 Social adjustment as a function of time

5.4.5 Changing Covariance Structures of Longitudinal Models

When doing any longitudinal analysis in *R*, the default setting is that the errors are independent with equal error variance. For example, the ϵ_{it}s in model (5.1) are assumed to be independent among all 405 women at all 3 time points, all with a common within-women standard deviation of σ. This assumption might be impractical since data collected longitudinally may be correlated with each other as a result of the time series autocorrelation.

To override this default assumption, we can change the error structure by using the *correlation* setting. A great way to adjust the covariance structure is to specify that the longitudinal data have an autoregressive error structure since autoregressive error structures can model data in which measurements taken closer together will have a stronger correlation than measurements taken farther apart.

R package *nlme* implemented three common options for autoregressive error structures for longitudinal data:

1. *corAR1* error structure is a first-order autoregressive error structure for time measured in fixed intervals.
2. *corCAR1* error structure is a first-order autoregressive error structure for time measured in varying intervals. For example, when different women are measured

at different time points. Participant 1 is measured at 1.4, 4.8, and 8.1 months, and participant 2 is measured at 1.2, 4.3, and 8.5 time points.
3. *corARMA* error structure is an error structure that incorporates both an autoregressive and a moving average component.

5.4.5.1 *R* Implementation for Random-Intercept Longitudinal Model

For random-intercept only model (such as in *mSoc2*) to examine *SocAdj* measurements over time, if we believe that the errors among the longitudinal measures follow an autoregressive process, we could use the following *R* code chunk to implement this autoregressive correlation:

```
# Add autoregressive error
mSoc2.2=lme(fixed=SocAdj~Time,random=~1|ID,
      correlation = corAR1(), data = dSoc)
# Print the summary of model fitting
summary(mSoc2.2)
```

```
## Linear mixed-effects model fit by REML
##   Data: dSoc
##        AIC      BIC    logLik
##    7387.01 7411.797 -3688.505
##
## Random effects:
##   Formula: ~1 | ID
##          (Intercept) Residual
## StdDev:    6.786224 6.892153
##
## Correlation Structure: AR(1)
##   Formula: ~1 | ID
##   Parameter estimate(s):
##        Phi
## 0.3270698
## Fixed effects: SocAdj ~ Time
##                 Value Std.Error  DF   t-value p-value
## (Intercept) 100.90924 0.5215996 666 193.46112  0.0000
## Time         -0.08999 0.0695569 666  -1.29381  0.1962
##   Correlation:
##        (Intr)
## Time -0.561
##
## Standardized Within-Group Residuals:
##          Min            Q1          Med          Q3          Max
## -4.996232468 -0.439554330 -0.002519174  0.438018220  4.482188774
##
## Number of Observations: 1053
## Number of Groups: 386
```

The results above indicate that the point estimate for the autocorrelation is *phi* = 0.327, which is quite large. For statistical testing, there are two paths that we can take:

1. to call *anova* to test this model *mSoc2.2* with autoregressive correlation to the model *mSoc2* without autoregressive correlation. The implementation is as follows:

```
# Test whether the autoregressive is significant
anova(update(mSoc2, method="ML"), update(mSoc2.2,method="ML"))
```

```
##
##                                   Model df      AIC      BIC
## update(mSoc2, method = "ML")          1  4 7393.454 7413.292
## update(mSoc2.2, method = "ML")        2  5 7383.672 7408.469
##
##                                   logLik    Test L.Ratio  p-value
## update(mSoc2, method = "ML")     -3692.727
## update(mSoc2.2, method = "ML")   -3686.836 1 vs 2 11.782   6e-04
```

2. to call the *R* function, *intervals*, to produce 95% confidence intervals for all estimated parameters. The implementation is as follows:

```
# Print the 95% CI for all parameter estimates
library(nlme)
intervals(mSoc2.2)
```

```
## Approximate 95% confidence intervals
##
##  Fixed effects:
##                     lower          est.          upper
## (Intercept) 99.8850596 100.90923720 101.93341480
## Time          -0.2265706  -0.08999335    0.04658394
## attr(,"label")
## [1] "Fixed effects:"
##
##  Random Effects:
##    Level: ID
##                       lower      est.    upper
## sd((Intercept)) 5.641577 6.786224 8.163114
##
##  Correlation structure:
##         lower      est.     upper
## Phi 0.1165188 0.3270698 0.509487
## attr(,"label")
## [1] "Correlation structure:"
##
##  Within-group standard error:
##    lower      est.    upper
## 5.960703 6.892153 7.969155
```

As seen from the *anova*, the *AIC* and *BIC* associated with *mSoc2.2* are smaller than those of *mSoc2*. The likelihood ratio test indicated that there is a statistically significant correlation existing since the *p-value=6e-4*. From the *intervals*, we can see that the 95% confidence interval is (0.117, 0.509), which does not include zero and indicates a statistically significant correlation with a correlation coefficient of 0.327. From a practical perspective, this result suggests that a positive correlation exists between adjacent pairs of *SocAdj* measurements—a relatively higher *SocAdj* at one time is associated with a relatively higher *SocAdj* at the next time point.

5.4.5.2   *R* Implementation for Random-Slope Longitudinal Model

Notice that when specifying error structures for a random-intercept only model, we use *correlation* $= cor AR1()$, which is in fact the short form of specifying the error structure as *correlation* $= cor AR1(form =\sim 1|ID)$. To model the error structure to a random-slope model, the random coefficients' structure must be specified *correlation* $= cor AR1(form =\sim Time|ID)$ in the *R* syntax as follows:

```
# Add autoregressive error to random-slope model
mSoc5.2=lme(fixed=SocAdj~Time*SurgTx, random = ~Time|ID,
            correlation=corAR1(form=~Time|ID),data=dSoc)
# Print the summary of model fitting
summary(mSoc5.2)
```

```
## Linear mixed-effects model fit by REML
##   Data: dSoc
##         AIC      BIC    logLik
##    7389.299 7433.899 -3685.65
##
## Random effects:
##  Formula: ~Time | ID
##  Structure: General positive-definite, Log-Cholesky parametrization
##              StdDev    Corr
## (Intercept) 8.6914250 (Intr)
## Time        0.5962565 -0.486
## Residual    5.4897019
##
## Correlation Structure: ARMA(1,0)
##  Formula: ~Time | ID
##  Parameter estimate(s):
## Phi1
##    0
## Fixed effects: SocAdj ~ Time * SurgTx
##                   Value Std.Error  DF  t-value p-value
## (Intercept) 103.73835 1.1894180 665 87.21775  0.0000
## Time         -0.43878 0.1481461 665 -2.96182  0.0032
## SurgTx       -3.58092 1.3336001 384 -2.68516  0.0076
## Time:SurgTx   0.43055 0.1663303 665  2.58853  0.0098
##  Correlation:
##
##               (Intr) Time   SurgTx
## Time        -0.598
## SurgTx      -0.892  0.533
## Time:SurgTx  0.533 -0.891 -0.597
##
```

```
## Standardized Within-Group Residuals:
##        Min          Q1         Med          Q3         Max
## -3.64987767 -0.43517395  0.00559863  0.42752515  4.54845478
##
## Number of Observations: 1053
## Number of Groups: 386
```

```
# Model comparison
anova(update(mSoc5,method="ML"), update(mSoc5.2,method="ML"))
```

```
##                                  Model df      AIC      BIC
## update(mSoc5, method = "ML")        1  8 7384.115 7423.790
## update(mSoc5.2, method = "ML")      2  9 7386.115 7430.749
##                                  logLik    Test      L.Ratio p-value
## update(mSoc5, method = "ML")   -3684.057
## update(mSoc5.2, method = "ML") -3684.057 1 vs 2 1.818989e-11 1
```

We can see that the estimated autoregressive correlation is zero (i.e., *Phi = 0*) and that testing *mSoc5* versus *mSoc5.2* is not statistically significant with a *p-value=1*. Therefore, *mSoc5* should be used.

For more possible covariance structures available in the *nlme* package, interested readers can search the *R* help using

```
?corClasses
```

5.5 Exercises

1. Follow the steps in this chapter and analyze the longitudinal data using *mood.*
2. Using the *National Longitudinal Survey of Freshmen* in Appendix A.6, analyze the longitudinal data from the 1051 African-American college students on their self-rated *academic effort* assessed during the second semester of their freshmen (Wave 2), sophomore (Wave 3), junior (Wave 4), and senior (Wave 5) years.
3. As a follow-up to the data analysis in Question 2, demonstrate the addition of *sex* and *major* (Science, Technology, Engineering, or Mathematics [STEM] versus Non-STEM) as possible predictors that may account for any change in the individual growth trajectories (i.e., intercept and slope) of self-rated academic effort.

Chapter 6
Nonlinear Regression Modeling

Not all models are linear. The truth is that there are more nonlinear models than linear models that exist in real-life applications. Although different from the standard linear regression that we have known, nonlinear regression can and should be used to model nonlinear relationships present in the real data. There are many different types of nonlinear models as discussed in Bates and Watts (1998). An example to model the nonlinear relationship between the proportion of available chlorine in chlorine solution and the length of time in weeks from a chlorine data using SAS/R/Stata/SPSS to illustrate the model fitting can be found in Chapter 10 of Wilson and Chen (2021).

To illustrate nonlinear multi-level modeling, we use this chapter to introduce the nonlinear regression first as the introduction for the next chapter on nonlinear multi-level modeling. In this chapter, we focus on a population dynamics model (i.e., population growth model), which is a typical nonlinear model.

6.1 Population Growth Models

Population growth modeling is one type of mathematical model that can be used to describe population dynamics. Such populations can be human populations, disease populations (such as COVID-19), or any ecological and environmental populations. For example, in ecological population modeling, we can focus on the changes in population parameters for population size and age distributions within a population (which may be due to interactions within the environment, individuals of their own species, or other species).

Historically, in the late eighteenth century, scientists and biologists began developing population modeling as a means of understanding population dynamics. In one of the first noted publications in this time period, Malthus (1798) described a geometric population growth model in his essay. Another well-known, milestone

© The Author(s), under exclusive license to Springer Nature Switzerland AG 2021
D.-G. (Din) Chen, J. K. Chen, *Statistical Regression Modeling with R*,
Emerging Topics in Statistics and Biostatistics,
https://doi.org/10.1007/978-3-030-67583-7_6

model was the logistic population growth by Verhulst (1838). In this logistic model, a sigmoid curve is used to describe population dynamics as exponential, followed by a decrease in growth, and finally asymptotically bound by a carrying capacity that is due to environmental limitations.

In this chapter, we will use the logistic growth model by Verhulst to illustrate the nonlinear model and nonlinear mixed-effects model in Chap. 7.

6.1.1 Model Development

Letting $N(t)$ represent the (cumulative) population size at time t, this logistic growth model can be formalized by the differential equation:

$$\frac{dN(t)}{dt} = r \times N(t) \times \left(1 - \frac{N(t)}{N_{max}}\right), \tag{6.1}$$

where r is the growth rate parameter and N_{max} is the carrying capacity parameter.

As seen from this model in Eq. (6.1), the population at the beginning is modeled by the first part of the equation $r \times N(t)$, where the value of rate r represents the proportional increase of the population $N(t)$ in one unit of time. As the population grows, the second part of $r \times N(t)^2/N_{max}$ increases to the amount almost as large as the first part, as more individuals in the population compete for limited resources in the critical carrying capacity N_{max}. This antagonistic effect is called the *bottleneck* in population dynamics and is modeled by the value of the parameter N_{max}. It is this competition that diminishes the exponential growth of the first part and results in the total population $N(t)$ ceasing to grow (this is called maturity of the population).

This differential equation (6.1) can be solved with the solution as follows:

$$N(t) = \frac{N_{max} N_0 e^{rt}}{N_{max} + N_0 (e^{rt} - 1)} = \frac{N_{max}}{1 + \left(\frac{N_{max} - N_0}{N_0}\right) e^{-rt}}, \tag{6.2}$$

where N_0 is the initial population at time 0. It can be shown that $\lim_{t \to \infty} N(t) = N_{max}$, which is to say that N_{max} is the limiting value of the population $N(t)$ as t approaches to infinity, i.e., the highest value that the population can reach given infinite time (or close to reaching in finite time).

To model the population at each unit time interval $Y(t)$ at time t, we can typically take the first derivative of equation (6.2). We can get

$$Y(t) = \frac{dN(t)}{dt} = N'(t) = \frac{r N_{max} \left(\frac{N_{max} - N_0}{N_0}\right) e^{-rt}}{\left\{1 + \left(\frac{N_{max} - N_0}{N_0}\right) e^{-rt}\right\}^2}, \tag{6.3}$$

which is always greater than 0, indicating that the population size is always growing from N_0 to N_{max} as time progresses. However, this growth is faster in the beginning and reaches a *tipping point* where population growth will slow down. To see this, let us take the second derivative of equation (6.2) (i.e., the first derivative of Eq. (6.3) to model the population change rate of $C(t)$). We can get

$$C(t) = \frac{d^2 N(t)}{dt^2} = N''(t)$$

$$= -\frac{r^2 N_{max} \left(\frac{N_{max}-N_0}{N_0}\right) e^{-rt}}{\left\{1 + \left(\frac{N_{max}-N_0}{N_0}\right) e^{-rt}\right\}^3} \left[1 - \left(\frac{N_{max}-N_0}{N_0}\right) e^{-rt}\right]. \quad (6.4)$$

Therefore, letting $\frac{d^2 N(t)}{dt^2} = 0$ in Eq. (6.4), we can get $t_{mid} = \frac{1}{r} ln\left(\frac{N_{max}-N_0}{N_0}\right)$, which is the so-called inflection point or tipping point when $t \le t_{mid}$, $\frac{d^2 N(t)}{dt^2} \ge 0$. The population increases faster as it approaches this tipping point and then slows down afterward, since after this tipping point $t \ge t_{mid}$ and $\frac{d^2 N(t)}{dt^2} \le 0$.

Using the tipping point of $t_{mid} = \frac{1}{r} ln\left(\frac{N_{max}-N_0}{N_0}\right)$, we can reformulate the logistic population model in Eq. (6.2) as follows:

$$N(t) = \frac{N_{max}}{1 + e^{-r(t-t_{mid})}}. \quad (6.5)$$

Model in Eq. (6.5) is the so-called three-parameter (i.e., N_{max}-maximum capacity, r-growth rate, and t_{mid}-tipping point) logistic growth curve model which is commonly used in modeling population growth. An extension to five-parameter logistic growth model in analyzing COVID-19 data in USA can be found in Chen et al. (2020).

Similarly, the population at unit time t in Eq. (6.3) can be obtained as follows:

$$Y(t) = N'(t) = \frac{r N_{max} e^{-r*(t-t_{mid})}}{\left\{1 + e^{-r(t-t_{mid})}\right\}} \quad (6.6)$$

and the population change rate $C(t)$ at time t in Eq. (6.4) as follows:

$$C(t) = N''(t) = -\frac{r^2 N_{max} e^{-r(t-t_{mid})}}{\left\{1 + e^{-r(t-t_{mid})}\right\}^3} [1 - exp(-r(t - t_{mid}))]. \quad (6.7)$$

To analyze real data with measurement errors, an error term is added to the three-parameter logistic growth model (6.5) to be used for statistical analysis as follows:

$$N(t) = \frac{N_{max}}{1 + e^{-r(t-t_{mid})}} + \epsilon(t), \quad (6.8)$$

where $\epsilon(t)$ is assumed to be random and is normally distributed with a mean of 0 and a standard deviation of σ, i.e., $\epsilon(t) \sim N(0, \sigma^2)$. The parameter σ in Eq. (6.8) is the so-called *within*-group variation.

6.1.2   Illustration of the Logistic Growth Model

Let us illustrate the above models. Assume a population spans from year 1 to 60 (i.e., $t = 1$ to 60). The initial population $N_0 = 1$ and the maximum can reach up to 5000 (i.e., $N_{max} = 5000$), with a population growth rate of $r = 10\%$ per year. The **R** code to generate the data is as follows:

```
# Time from the beginning of Year 0 to Year 60
tmp = 0:60
# Population parameters
N0 = 1; Nmax = 5000; r = 0.30
# Calculate the tipping point
tmid = log((Nmax-N0)/N0)/r;tmid
```

```
## [1] 28.38998
```

```
# Calculate the cumulative population size: N(t)
Nt = Nmax/(1+exp(-(tmp-tmid)));
# Print the cumulative population size
round(Nt,2)
```

```
##  [1]    0.00    0.00    0.00    0.00    0.00    0.00    0.00
##  [8]    0.00    0.00    0.00    0.00    0.00    0.00    0.00
## [15]    0.00    0.01    0.02    0.06    0.15    0.42    1.14
## [22]    3.09    8.38   22.71   61.25  163.05  419.70  997.06
## [29] 2018.61 3239.73 4167.07 4657.52 4868.31 4950.73 4981.76
## [36] 4993.28 4997.52 4999.09 4999.66 4999.88 4999.95 4999.98
## [43] 4999.99 5000.00 5000.00 5000.00 5000.00 5000.00 5000.00
## [50] 5000.00 5000.00 5000.00 5000.00 5000.00 5000.00 5000.00
## [57] 5000.00 5000.00 5000.00 5000.00 5000.00
```

```
# Calculate the yearly population size: Y(t) = N'(t)
Yt = r*Nmax*exp(-r*(tmp-tmid))/(1+exp(-r*(tmp-tmid)))^2;
# Print the yearly population size
round(Yt,2)
```

```
##  [1]    0.30    0.41    0.55    0.74    1.00    1.34    1.82    2.45
##  [9]    3.31    4.46    6.03    8.14   10.98   14.82   20.01   27.00
## [17]   36.44   49.16   66.31   89.36  120.27  161.49  215.90  286.46
## [25]  374.99  479.76  591.37  689.21  744.90  737.61  670.28  567.09
## [33]  455.63  353.97  269.42  202.64  151.39  112.68   83.69   62.08
```

```
## [41]   46.03   34.11   25.28   18.73   13.88   10.28    7.62    5.64
## [49]    4.18    3.10    2.29    1.70    1.26    0.93    0.69    0.51
## [57]    0.38    0.28    0.21    0.15    0.11
```

```
# Calculate the population change rate: C(t)= N''(t)
Ct = -r^2*Nmax*exp(-r*(tmp-tmid))*(1-exp(-r*(tmp-tmid)))/
(1+exp(-r*(tmp-tmid)))^3;
# Print the population change
round(Ct,2)
```

```
##  [1]    0.09    0.12    0.16    0.22    0.30    0.40    0.54    0.73
##  [9]    0.98    1.32    1.78    2.39    3.20    4.28    5.69    7.54
## [17]    9.93   12.96   16.73   21.25   26.45   32.03   37.40   41.59
## [25]   43.30   41.15   34.11   22.15    6.55  -10.18  -25.16  -36.10
## [33]  -42.02  -43.21  -40.83  -36.28  -30.80  -25.26  -20.19  -15.83
## [41]  -12.23   -9.35   -7.09   -5.35   -4.01   -3.00   -2.24   -1.67
## [49]   -1.24   -0.92   -0.68   -0.51   -0.38   -0.28   -0.21   -0.15
## [57]   -0.11   -0.08   -0.06   -0.05   -0.03
```

With the data generated, we can plot these models. The population growth model in Eq. (6.5) can be shown in Fig. 6.1 as follows:

```
par(mar = c(4, 4, .1, .1))
plot(tmp, Nt, xlab="Time", ylab="Cumulative Population", type = 'b', pch = 19)
```

This is the so-called sigmoid growth curve where the population grows from $N_0 = 1$ to $N_{max} = 5000$. The yearly population curve in model (6.6) can be shown in Fig. 6.2 as follows:

```
par(mar = c(4, 4, .1, .1))
plot(tmp, Yt, xlab="Time", ylab="Yearly Population", type = 'b', pch = 19)
```

From this model, the yearly population as described in model (6.6) will grow yearly to the tipping point of year $t_{mid} = 28.39$. Then, the population will decline to zero as seen in Fig. 6.2. The yearly rate of change as described in model (6.7) will be positive before the tipping point and then negative afterward as seen in Fig. 6.3.

```
par(mar = c(4, 4, .1, .1))
plot(tmp, Ct, xlab="Time", ylab="Change Rate", type = 'b', pch = 19)
```

6.2   Theory of Nonlinear Regression with *nls*

6.2.1   R Packages for Nonlinear Regression

The nonlinear regression model is a generalization of the linear regression model, where the conditional mean of the response variable is a *nonlinear* function of the

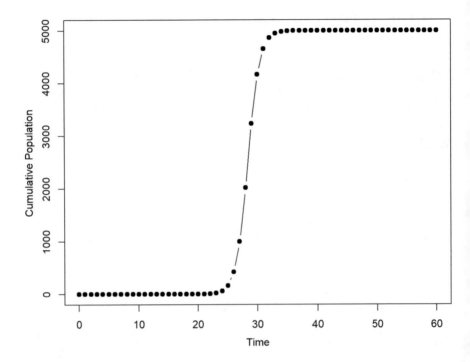

Fig. 6.1 Illustration of cumulative logistic growth model

parameters instead of a linear function of the parameters. As seen in Fig. 6.5, the US population (in millions) displays a nonlinear growth function. A linear regression would not be appropriate in this case and would not fit the data. A nonlinear model should be used. There are many nonlinear models that can be used to model this type of nonlinear growth—the logistic growth model in Eq. (6.8) is one of the most popular models for population growth.

There are several R packages that can be used for nonlinear regression, to list a few:

- *Stats*: a collection of statistical functions by R Core Team and contributors worldwide;
- *nlreg*: Higher Order Inference for Nonlinear Heteroscedastic Models;
- *nls2*: nonlinear regression with brute force;
- *nlsem*: estimation of structural equation models with nonlinear effects and underlying non-normal distributions;
- *nlsMicrobio*: nonlinear regression in predictive microbiology;
- *nlstools*: tools for nonlinear regression analysis;
- *nnls*: the Lawson–Hanson algorithm for non-negative least squares (NNLS).

Interested readers can refer to Nash (2014) for details on other R tools for nonlinear modeling.

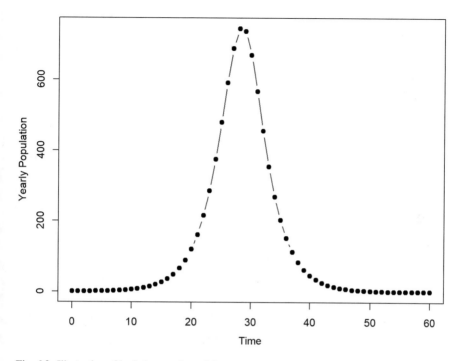

Fig. 6.2 Illustration of logistic growth model

For general use, the *stats* package has almost all the functionalities needed. We will use the *nls* function in this package to describe the nonlinear regression. The *nls* is used to determine the nonlinear (weighted) least squares estimates of the parameters of a nonlinear model (Bates and Watts 1998). The general use of *nls* is as follows:

```
nls(formula, data, start, control, algorithm,
    trace, subset, weights, na.action, model,lower, upper, ...)
```

where

- *formula* is a nonlinear model formula including variables and parameters,
- *data* is an optional dataframe in which to evaluate the variables in formula and weights,
- *start* is a list of starting estimates,
- *control* is an optional list of control settings for the optimization,
- *algorithm* is character string specifying the algorithm to use,
- *trace* is a logical value indicating if a trace of the iteration progress should be printed,
- *subset* is an optional vector specifying a subset of data to be used in the fitting process,

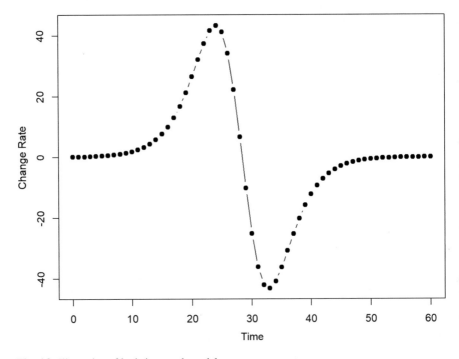

Fig. 6.3 Illustration of logistic growth model

- *weights* is an optional numeric vector of known weights for the weighted least squares,
- *na.action* is a function on what should happen when the data contains NAs,
- *model* is a *TRUE/FALSE* logical on whether or not to return the model frame,
- *lower* and *upper* are vectors of lower and upper bounds.

6.2.2 *Theory of Nonlinear Regression in* nls

For a general description of nonlinear regression, let us denote y_i (as the US population at decade i) and x_i (as the time in decade) as the response and independent predictors at their ith observation (where $i = 1, \cdots, n$). Then, the general nonlinear model is described as follows:

$$y_i = E(y_i|x_i) = f(x_i, \theta) + \epsilon_i. \tag{6.9}$$

In this model, the mean response of $E(y|x)$ depends on x through the mean function $f(x, \theta)$, where the predictor x can have one or more components, and the parameter vector θ can also have one or more components. In the logistic growth model in

Eq. (6.8), x consists of the single predictor *decade*, and the parameter vector $\theta = c(N_{max}, t_{mid}, r)$ has three components.

If $f(x, \theta) = \beta_0 + \beta_1 x_1 + \cdots + \beta_p x_p$, where $\theta = (\beta_0, \beta_1, \cdots, \beta_p)$, the general nonlinear regression model in Eq. (6.9) becomes the general multiple linear regression. In this sense, the multiple linear regression model is a very special case of the general nonlinear regression model.

In the model (6.9), we further assume that the errors ϵ_i are independently distributed with error variance parameter σ^2/w_i, where σ^2 is an unknown parameter representing *within*-group variation, but w_is are assumed to be known as non-negative weights.

In the R package *stats*, the *nls* function is to estimate the parameter vector θ that minimizes the residual sum of squares,

$$RSS(\theta) = \sum_{i=1}^{n} w_i \, [y_i - f(x_i, \theta)]^2 . \qquad (6.10)$$

We denote the estimated θ as $\hat{\theta}$. Note that the weighted nonlinear regression would be simplified to the typical nonlinear regression if $w_i = 1$.

Different from the general linear regression with least squares estimation in Chap. 1, there may not be a closed-form analytical formula to calculate the $\hat{\theta}$ depending on the nonlinear model $f(x, \theta)$. In those cases, an iterative optimization procedure must be used to obtain the parameter estimates. Therefore, a good initial value is very helpful to search for the optimal estimate of $\hat{\theta}$ in (6.10). This is especially critical if the *RSS* in Eq. (6.10) is not concave with multiple local minima. The default algorithm is a *Gauss–Newton algorithm*. Other choices are *plinear* for the *Golub–Pereyra algorithm* for partially linear least squares models and *port* for the *nl2sol algorithm* from the *Port* library.

6.3 Nonlinear Regression for US Population Data

6.3.1 Data on US Population

Let us start with the US population data described in Sect. A.7 in Appendix A. Let us read the data into the R session as follows:

```
# Read in the US population data from 1610 to 2010
dUSA = read.csv("uspopulation.csv", header=T)
## print the summary of the data
# Scale the population in millions
dUSA$sPop = dUSA$Pop/1000000
# Print the data summary
summary(dUSA)
```

```
##      Decade            Pop                  Grate          sPop
##  Min.   :1610   Min.   :        350   Min.   :0.0730   Min.   :   0.00035
##  1st Qu.:1710   1st Qu.:     331711   1st Qu.:0.1910   1st Qu.:   0.33171
##  Median :1810   Median :    7239881   Median :0.3290   Median :   7.23988
##  Mean   :1810   Mean   :   58596031   Mean   :0.5569   Mean   :  58.59603
##  3rd Qu.:1910   3rd Qu.:   92228496   3rd Qu.:0.3703   3rd Qu.:  92.22850
##  Max.   :2010   Max.   :  308745538   Max.   :5.5770   Max.   : 308.74554
##                                       NA's   :1
```

```
# Print the first 10 observations
head(dUSA, 10)
```

```
##    Decade    Pop Grate      sPop
## 1    1610    350    NA 0.000350
## 2    1620   2302 5.577 0.002302
## 3    1630   4646 1.018 0.004646
## 4    1640  26634 4.733 0.026634
## 5    1650  50368 0.891 0.050368
## 6    1660  75058 0.490 0.075058
## 7    1670 111935 0.491 0.111935
## 8    1680 151507 0.354 0.151507
## 9    1690 210372 0.389 0.210372
## 10   1700 250888 0.193 0.250888
```

We can plot the population growth rate with the following *R* code chunk:

```
# Plot the US population growth rates
par(mar = c(4, 4, .1, .1))
plot(Grate~Decade, xlab="Decade", ylab="Population Growth Rate",
     data = dUSA, type = 'b', pch = 19)
```

As seen from Fig. 6.4, the population growth rates have been declining from more than 100% in the early seventeenth century (i.e., 557.7% in 1620s, 101.8% in 1630s, 473.3% in 1640s, 89.1% in 1950s, etc.) to about 10% in the recent century (i.e., 9.8% in 1990s, 13.2% in 2000s, and 9.7% in 2010s). We can also plot the population size using the following *R* code chunk:

```
# Make the plot on population growth
par(mar = c(4, 4, .1, .1))
plot(sPop~Decade, xlab="Decade", ylab="Total Population (in Millions)",
     data = dUSA, type = 'b', pch = 19)
```

It can be seen from Fig. 6.5 that the US population grew from a few hundreds in the early seventeenth century to 300 million in this century in a nonlinear fashion. A linear model would not be appropriate to describe this growth model, so a nonlinear regression model should be used.

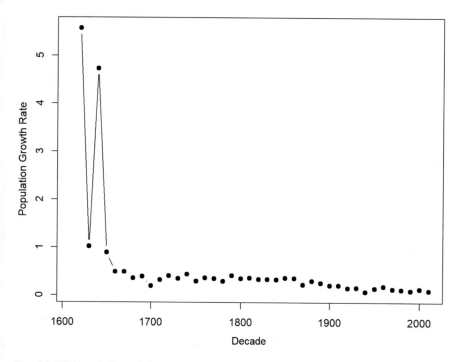

Fig. 6.4 US historical population

6.3.2   Fitting the US Population Data Using nls

Let us fit the data *dUSA* to the nonlinear logistic growth model in Eq. (6.5). Since
we need a set of good initial values for *nls* to converge, we make an educated
guesstimate that a set of initial values could be *Nmax = 1000, tmid = 1950, and
r = 0.2* based on Fig. 6.5, which means that we assume that the maximum carrying
capacity in the USA can go to 1,000 million (i.e., *Nmax=1000*), with a population
growth rate of 20% (i.e., *r=0.2*) and a tipping point at about the year 1950 (i.e.,
tmid=1950). With this set of parameters, we can run *nls* as follows:

```
# Use the default Gauss-Newton search algorithm
pop.mod = nls(sPop~Nmax/(1+exp(-r*(Decade-tmid))), data=dUSA,
          start=list(Nmax=1000, tmid=1950,r=0.2), trace=T)

## 3418999 :   1000.0 1950.0      0.2
## 48728.86 :    267.6889075 1949.4015554      0.1027154
## 12497.95 :  2.831193e+02 1.948585e+03 2.719032e-02
## 4208.574 :   414.1620733 1976.0349999      0.0187547
## 710.0094 :  4.717263e+02 1.984022e+03 2.114019e-02
## 570.5819 :  4.747903e+02 1.983043e+03 2.111299e-02
## 570.5513 :  4.737365e+02 1.982827e+03 2.114289e-02
```

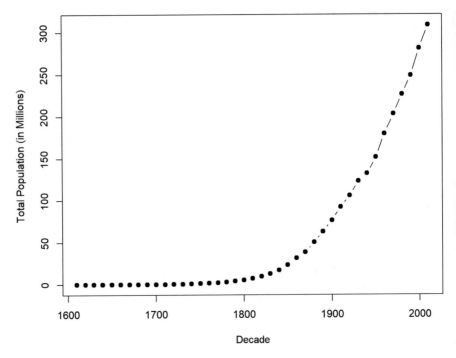

Fig. 6.5 US historical population

```
## 570.551 :    4.738039e+02 1.982842e+03 2.114015e-02
## 570.551 :    4.737937e+02 1.982839e+03 2.114051e-02
## 570.551 :    4.737948e+02 1.982840e+03 2.114047e-02
```

```
#Print the model fit
summary(pop.mod)
```

```
##
## Formula: sPop ~ Nmax/(1 + exp(-r * (Decade - tmid)))
##
## Parameters:
##          Estimate Std. Error t value Pr(>|t|)
## Nmax 4.738e+02  2.433e+01   19.48   <2e-16 ***
## tmid 1.983e+03  5.055e+00  392.22   <2e-16 ***
## r    2.114e-02  6.605e-04   32.01   <2e-16 ***
## ---
## Signif. codes:  0 '***' 0.001 '**' 0.01 '*' 0.05 '.' 0.1 ' ' 1
##
## Residual standard error: 3.875 on 38 degrees of freedom
##
## Number of iterations to convergence: 9
## Achieved convergence tolerance: 1.38e-06
```

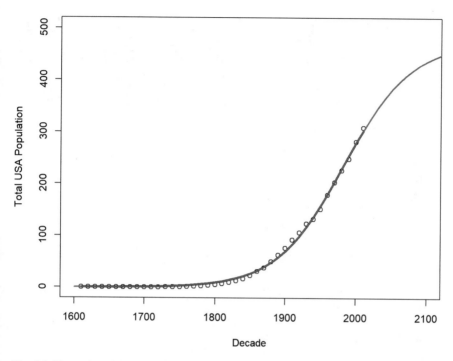

Fig. 6.6 Illustration of US population growth model

We obtained a converged solution. With this model, the estimated carrying capacity in the USA is $\tilde{N}_{max} = 473.8$ million with a population growth rate of $\hat{r} = 2.114\%$. The tipping point (i.e., change point) is estimated at the year $\hat{t}_{mid} = 1983$.

Note that if we make an *educated guess* that the initial value is $r = 0.1$ or $r = 0.3$, we will have the same converged solution. However, if we *guesstimated* that the initial value was $r = 0.4$, the *nls* would crash and no solution would be able to be found. Interested readers can try a few initial *guestimates* to see whether you can get the same final estimates.

With this satisfactory model fitting, we can predict the population and overlay the predicted population with the observed population. The following *R* code chunk can be used to display this illustration (Fig. 6.6).

```
par(mar = c(4, 4, .1, .1))
# plot the data and model fit
plot(sPop ~ Decade, xlim= c(1600,2100), ylim=c(0, 500),data=dUSA,
     xlab="Decade", ylab="Total USA Population")

# add the model predicted
lines(dUSA$Decade, predict(pop.mod), lwd=3)

# Population Prediction to 2150 based on the model
```

```
newY = data.frame(Decade = seq(1600, 2150, by=10))
pred.pop = predict(pop.mod, newY);
lines(newY$Decade, pred.pop, col="red", lwd=2)
```

```
# Print the fitted and predicted population
print(data.frame(Decade=newY$Decade, new.sPop=round(pred.pop,1)))
```

```
##      Decade new.sPop
## 1     1600      0.1
## 2     1610      0.2
## 3     1620      0.2
## 4     1630      0.3
## 5     1640      0.3
## 6     1650      0.4
## 7     1660      0.5
## 8     1670      0.6
## 9     1680      0.8
## 10    1690      1.0
## 11    1700      1.2
## 12    1710      1.5
## 13    1720      1.8
## 14    1730      2.2
## 15    1740      2.8
## 16    1750      3.4
## 17    1760      4.2
## 18    1770      5.2
## 19    1780      6.4
## 20    1790      7.9
## 21    1800      9.7
## 22    1810     12.0
## 23    1820     14.7
## 24    1830     18.0
## 25    1840     22.1
## 26    1850     26.9
## 27    1860     32.9
## 28    1870     39.9
## 29    1880     48.4
## 30    1890     58.4
## 31    1900     70.1
## 32    1910     83.7
## 33    1920     99.2
## 34    1930    116.8
## 35    1940    136.4
## 36    1950    157.8
## 37    1960    180.8
```

```
## 38    1970    204.9
## 39    1980    229.8
## 40    1990    254.8
## 41    2000    279.4
## 42    2010    303.1
## 43    2020    325.4
## 44    2030    346.1
## 45    2040    364.8
## 46    2050    381.6
## 47    2060    396.3
## 48    2070    409.0
## 49    2080    419.9
## 50    2090    429.2
## 51    2100    437.1
## 52    2110    443.6
## 53    2120    449.1
## 54    2130    453.6
## 55    2140    457.3
## 56    2150    460.4
```

With this model, we predict the US population size to be 325.44 million in the Census of 2020.

6.4 Exercises

The *USPop* data is a decennial time series of the US population from 1790 to 2000, compiled in Fox and Weisberg (2016):

1. Use the data available from *R* package *car* (i.e., *Companion to Applied Regression*) (Fox and Weisberg 2019) and practice the model fitting in this chapter.
2. Compare the estimates from this data with the estimates obtained from this chapter.
3. Explain the differences between the Exercises 1 and 2 above.

Using the *State of North Carolina COVID-19 Data* in Appendix A.8:
4. Fit the logistic growth model in Eq. (6.8) and report your findings.

Chapter 7
Nonlinear Mixed-Effects Modeling

Continuing from Chap. 6, this chapter illustrates the nonlinear MLM to estimate the additional *between*-group variation in addition to the *within*-group variation discussed in Chap. 6. For a detailed theory of nonlinear MLM (i.e., nonlinear mixed-effects model), interested readers can refer to the book by (Pinheiro and Bates 2000). In this chapter, we illustrate the application of the *R* package *nlme* by analyzing a commonly used dataset on the growth of loblolly pines.

Note to readers: We will use *R* package *nlme* (i.e., *nonlinear mixed-effects* model) in this chapter. Remember to install this *R* package to your computer using *install.packages("nlme")* and load this package into *R* session using *library(nlme)* before running all the *R* programs in this chapter.

7.1 Dataset of *Loblolly* Pine Trees' Growth

The loblolly pine is a type of pine commonly seen in the Southeastern United States from central Texas, to eastern Florida, and all the way up to northern Delaware and southern New Jersey. According to the US Forest Service, the loblolly pine is the second most common species of tree in the United States after red maple. For timber production, the pine species is regarded as the most commercially important tree in the Southeastern United States.

A dataset named *Loblolly* from 14 loblolly pine trees are available (Pinheiro and Bates 2000). This dataset is a dataframe of 84 rows and 3 columns of the growth of loblolly pine trees which is built in the *R* package *nlme*. The dataset includes 3 variables:

- *height*: a numeric vector of tree heights (ft),
- *age*: a numeric vector of tree ages (yr) from year 3 to year 25, and

© The Author(s), under exclusive license to Springer Nature Switzerland AG 2021
D.-G. (Din) Chen, J. K. Chen, *Statistical Regression Modeling with R*,
Emerging Topics in Statistics and Biostatistics,
https://doi.org/10.1007/978-3-030-67583-7_7

- *Seed*: an ordered factor indicating the seed source for the tree. The ordering is by increasing maximum height.

Let us load the *R* library and examine the data as follows:

```
# Load the R library
library(nlme)
# Load the data
data(Loblolly)
# Print the first part of the data
head(Loblolly)
```

```
## Grouped Data: height ~ age | Seed
##       height age Seed
## 1      4.51   3  301
## 15    10.89   5  301
## 29    28.72  10  301
## 43    41.74  15  301
## 57    52.70  20  301
## 71    60.92  25  301
```

We can plot the data as follows:

```
# Load the R library
# Plot the data
plot(Loblolly)
```

In Fig. 7.1, the points are the observed tree heights in feet, and the solid line is to link the observed heights. It can be seen from this figure that the 14 trees' growth trajectories are nonlinear.

7.2   Nonlinear Regression for *Loblolly* Growth

To further demonstrate nonlinear regression from Chap. 6, we can make use of the nonlinear logistic growth model (6.8) in Chap. 6 again to illustrate the nonlinear regression. To describe the height of the loblolly tree, let us rewrite this logistic growth model with different notation as follows:

$$h(t) = \frac{H_{max}}{1 + e^{-r(t - t_{mid})}} + \epsilon(t), \tag{7.1}$$

where $h(t)$ describes the height of a loblolly pine at its tree *age* of t years ($t = 3$, ...,25), r is still the parameter of growth rate, and t_{mid} is the tipping point year when the loblolly tree reaches its middle point of growth. ϵ is assumed to be random and is

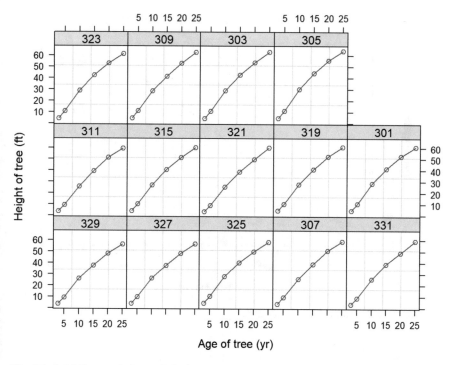

Fig. 7.1 Loblolly growth data and plotting

normally distributed with mean of 0 and standard deviation of σ, i.e., $\epsilon \sim N(0, \sigma^2)$. The parameter σ is the so-called *within*-group (i.e., trees) variation.

This nonlinear regression model in Eq. (7.1) is designed to analyze one *unit* (i.e., one tree in this data). We illustrate this by subset the dataset to use the *seed=329*, which can be seen in Fig. 7.2, as follows:

```
# Plot the growth curve of tree #329
plot(height ~ age, data = Loblolly, subset = Seed == 329,
     xlab = "Tree age (yr)", las = 1,
     ylab = "Tree height (ft)",
     main = "Loblolly Data and Fitted Model (Seed 329 only)")

# fit a nonlinear regression
fm329 = nls(height ~ Hmax/(1+exp(-r*(age-tmid))),
            start= c(Hmax=60, tmid = 15,r=0.1),
               data = Loblolly, subset = Seed == 329)
summary(fm329)
```

```
##
## Formula: height ~ Hmax/(1 + exp(-r * (age - tmid)))
##
## Parameters:
```

Loblolly Data and Fitted Model (Seed 329 only)

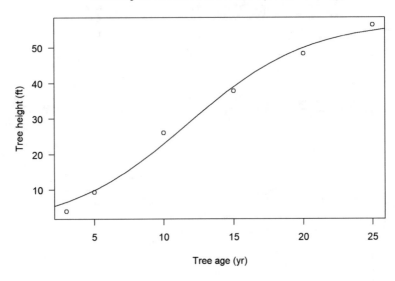

Fig. 7.2 Loblolly growth nonlinear regression

```
##        Estimate Std. Error t value Pr(>|t|)
## Hmax 57.33718    3.92615   14.604 0.000696 ***
## tmid 11.73869    1.02447   11.458 0.001427 **
## r     0.23296    0.03961    5.881 0.009811 **
## ---
## Signif. codes:  0 '***' 0.001 '**' 0.01 '*' 0.05 '.' 0.1 ' ' 1
##
## Residual standard error: 2.859 on 3 degrees of freedom
##
## Number of iterations to convergence: 8
## Achieved convergence tolerance: 6.099e-06
```

```
# Add the fitting
age <- seq(0, 30, length.out = 30)
lines(age, predict(fm329, list(age = age)))
```

From this model fitting, we can see that the estimated $\hat{H}_{max} = 57.337$ (ft), $\hat{t}_{mid} = 11.738$(yr), and the growth rate $\hat{r} = 0.232$. In addition, the estimated *within*-tree variation is $\hat{\sigma} = 2.859$.

7.3 Nonlinear MLM with *R* Package *nlme*

7.3.1 Nonlinear MLM

Nonlinear MLM is called nonlinear mixed-effects modeling. To incorporate the *between*-group (i.e., tree) variation among the 14 trees, model (7.1) can be extended as a nonlinear mixed-effects logistic growth model, which is a two-level mixed-effects model. The level 1 model is to describe the relationship between time and growth at each tree and can be denoted as follows:

$$\text{Level 1: } h_i(t) = \frac{H_{max,i}}{1 + e^{-r_i(t - t_{mid,i})}} + \epsilon_i(t), \tag{7.2}$$

where subscript i describes the ith tree. In this level 1 model in (7.2), the three growth parameters (H_{max}, r, and t_{mid}) are ith tree-specific.

To model the *between*-tree variation among the 14 trees, a level 2 model is specified as the random effects for ($H_{max,i}$, r_i, $t_{mid,i}$) to model the group effect as follows:

$$\text{Level 2: } H_{max,i} = H_{max} + \nu_{H_{max,i}}$$

$$r_i = r + \nu_{r_i}$$

$$t_{mid,i} = H_{mid} + \nu_{t_{mid,i}}, \tag{7.3}$$

where (H_{max}, r, t_{mid}) are the fixed effects to describe the overall group effects from these 14 trees. The νs are the level 2 random errors associated with the three parameters from each tree to the overall group effects (H_{max}, r, t_{mid}), which is assumed to be multivariate (i.e., 3) normally distributed as

$$\begin{pmatrix} \nu_{H_{max,i}} \\ \nu_{r_i} \\ \nu_{t_{mid,i}} \end{pmatrix} \sim N \left(\begin{pmatrix} 0 \\ 0 \\ 0 \end{pmatrix}, \begin{pmatrix} \tau^2_{H_{max}} & \tau_{H_{max}} \tau_r \rho_{H_{max},r} & \tau_{H_{max}} \tau_{t_{mid}} \rho_{H_{max},t_{mid}} \\ \tau_{H_{max}} \tau_r \rho_{H_{max},r} & \tau^2_r & \tau_r \tau_{t_{mid}} \rho_{r,t_{mid}} \\ \tau_{H_{max}} \tau_{t_{mid}} \rho_{H_{max},t_{mid}} & \tau_r \tau_{t_{mid}} \rho_{r,t_{mid}} & \tau^2_{t_{mid}} \end{pmatrix} \right),$$

where ($\tau_{H_{max}}$, τ_r, $\tau_{t_{mid}}$) are the standard deviations for the random effects to describe the between-tree variations, and ($\rho_{H_{max},r}$, $\rho_{H_{max},t_{mid}}$, $\rho_{r,t_{mid}}$) are the correlations between the three random effects.

Of course, the three random effects at model (7.3) can be reduced to two or one random effect(s) depending on the variation in the data. We will show this simplification in model fitting and model selection in upcoming sections.

7.3.2 R *Package* nlme

To analyze all 14 trees together to estimate the overall growth, we make use of the
R function *nlme* in package *nlme* for this illustration. The typical usage of *nlme*
function is

```
nlme(model, data, fixed, random, groups, start, correlation, weights,
     subset, method, na.action, naPattern, control, verbose)
```

where

- *model* is the nonlinear model formula,
- *data* is the dataframe containing the variables used in the *model*,
- *fixed* is a two-sided linear formula for the fixed effects,
- *random* is a two-sided linear formula for the random effects,
- *groups* is an optional one-sided formula specifying the partitions of the data over
  which the random effects vary,
- *start* is an optional numeric vector to list the initial estimates for the fixed effects
  and random effects,
- *correlation* is an optional *corStruct* object describing the within-group correlation structure,
- *weights* is an optional *varFunc* object or one-sided formula describing the within-
  group heteroscedasticity structure.
- *subset* is an optional expression indicating the subset of data to be used in the fit,
- *method* is a character string. If *REML*, the model is fit by maximizing the
  restricted log-likelihood, and if *ML*, the log-likelihood is maximized; defaults
  to "ML,"
- *na.action* is a function that indicates what should happen when the data contain
  NAs,
- *naPattern* is an expression or formula object, specifying which returned values
  are to be regarded as missing,
- *control* is a list of control values for the estimation algorithm, and
- *verbose* is an optional logical value. If *TRUE*, the information on the evolution
  of the iterative algorithm is printed. Default is FALSE.

7.3.3 Model Fitting

Using *nlme*, we illustrate the model fitting and model selection. For model fitting,
we illustrate a step-by-step implementation on a series of models from the most
general model in (7.3), where all 3 parameters are random effects, to the simplest
model where only the H_{max} is a random effect. For model selection, we make use
of R function *anova*, which will produce the likelihood ratio test statistics to select
the most parsimonious model for the data.

The R code for model fitting is as follows:

```
# First load the library
library(nlme)
# Model 1: Random-effects for all three parameters
Model1 <- nlme(height ~ Hmax/(1+exp(-r*(age-tmid))),
        start  = c(Hmax=60,tmid=11,r=0.2),
        data   = Loblolly,
        fixed  = Hmax+tmid+r ~ 1,
        random = Hmax+tmid+r ~ 1)
summary(Model1)
```

```
## Nonlinear mixed-effects model fit by maximum likelihood
##   Model: height ~ Hmax/(1 + exp(-r * (age - tmid)))
##  Data: Loblolly
##        AIC       BIC    logLik
##    412.5837 436.8919 -196.2918
##
## Random effects:
##  Formula: list(Hmax ~ 1, tmid ~ 1, r ~ 1)
##  Level: Seed
##  Structure: General positive-definite, Log-Cholesky parametrization
##          StdDev       Corr
## Hmax     1.950539e+00 Hmax    tmid
## tmid     7.350794e-05 -0.002
## r        4.272640e-08  0.000  0.000
## Residual 2.306948e+00
##
## Fixed effects: Hmax + tmid + r ~ 1
##          Value Std.Error DF  t-value p-value
## Hmax 61.23769 1.0139526 68 60.39502       0
## tmid 11.75006 0.2110242 68 55.68108       0
## r     0.23272 0.0081380 68 28.59675       0
##  Correlation:
##      Hmax    tmid
## tmid  0.717
## r    -0.633 -0.649
##
## Standardized Within-Group Residuals:
##        Min         Q1        Med        Q3       Max
## -1.5018780 -0.9051631 -0.2798603 0.6790280 1.9592586
##
## Number of Observations: 84
## Number of Groups: 14
```

```
# Model 2: Random-effects for Hmax and tmid
Model2 <- nlme(height ~ Hmax/(1+exp(-r*(age-tmid))),
    start  = c(Hmax=60,tmid=11,r=0.2),
    data   = Loblolly,
    fixed  = Hmax+tmid+r ~ 1,
    random = Hmax+tmid ~ 1)
summary(Model2)
```

```
## Nonlinear mixed-effects model fit by maximum likelihood
##   Model: height ~ Hmax/(1 + exp(-r * (age - tmid)))
##  Data: Loblolly
```

```
##        AIC       BIC    logLik
##   406.5837  423.5994  -196.2918
##
## Random effects:
##  Formula: list(Hmax ~ 1, tmid ~ 1)
##  Level: Seed
##  Structure: General positive-definite, Log-Cholesky parametrization
##           StdDev       Corr
## Hmax     1.950539e+00 Hmax
## tmid     6.893274e-05 -0.002
## Residual 2.306948e+00
##
## Fixed effects: Hmax + tmid + r ~ 1
##          Value Std.Error DF  t-value p-value
## Hmax 61.23769 1.0139526 68 60.39502       0
## tmid 11.75006 0.2110242 68 55.68108       0
## r     0.23272 0.0081380 68 28.59675       0
##  Correlation:
##       Hmax    tmid
## tmid   0.717
## r     -0.633 -0.649
##
## Standardized Within-Group Residuals:
##         Min         Q1        Med        Q3        Max
## -1.5018780 -0.9051631 -0.2798603 0.6790280 1.9592586
##
## Number of Observations: 84
## Number of Groups: 14
```

```
# Model 3: Random-effects for Hmax and r
Model3 <- nlme(height ~ Hmax/(1+exp(-r*(age-tmid))),
    start = c(Hmax=60,tmid=11,r=0.2),
    data  = Loblolly,
    fixed = Hmax+tmid+r ~ 1,
    random = Hmax+r ~ 1)
summary(Model3)
```

```
## Nonlinear mixed-effects model fit by maximum likelihood
##   Model: height ~ Hmax/(1 + exp(-r * (age - tmid)))
##  Data: Loblolly
##        AIC       BIC    logLik
##   406.5837  423.5994  -196.2918
##
## Random effects:
##  Formula: list(Hmax ~ 1, r ~ 1)
##  Level: Seed
##  Structure: General positive-definite, Log-Cholesky parametrization
##           StdDev       Corr
## Hmax     1.950539e+00 Hmax
## r        1.436119e-06 -0.001
## Residual 2.306948e+00
##
## Fixed effects: Hmax + tmid + r ~ 1
##          Value Std.Error DF  t-value p-value
## Hmax 61.23769 1.0139527 68 60.39501       0
```

```
## tmid 11.75006 0.2110242 68 55.68108       0
## r      0.23272 0.0081380 68 28.59675       0
##   Correlation:
##        Hmax    tmid
## tmid  0.717
## r    -0.633 -0.649
##
## Standardized Within-Group Residuals:
##         Min          Q1         Med          Q3         Max
## -1.5018780 -0.9051632 -0.2798603  0.6790280  1.9592588
##
## Number of Observations: 84
## Number of Groups: 14
```

```r
# Model 4: Random-effects for Hmax only
Model4 <- nlme(height ~ Hmax/(1+exp(-r*(age-tmid))),
    start = c(Hmax=60,tmid=11,r=0.2),
    data  = Loblolly,
    fixed = Hmax+tmid+r ~ 1,
    random = Hmax ~ 1)
summary(Model4)
```

```
## Nonlinear mixed-effects model fit by maximum likelihood
##   Model: height ~ Hmax/(1 + exp(-r * (age - tmid)))
##   Data: Loblolly
##           AIC       BIC     logLik
##     402.5837 414.7378 -196.2918
##
## Random effects:
##   Formula: Hmax ~ 1 | Seed
##                Hmax Residual
## StdDev: 1.950539 2.306948
##
## Fixed effects: Hmax + tmid + r ~ 1
##            Value Std.Error DF  t-value p-value
## Hmax 61.23769 1.0139527 68 60.39501       0
## tmid 11.75006 0.2110242 68 55.68108       0
## r     0.23272 0.0081380 68 28.59675       0
##
## Number of Observations: 84
## Number of Groups: 14
```

We fitted 4 models as seen above. These models are nested from the most general
model in *Model 1* to the simplest model in *Model 4*. With nested models, we can
use the likelihood ratio test to test model significance for model selection. We make
use of the *R* function *anova* as follows:

```r
# Model 1 to Model 2
anova(update(Model1,method="ML"), update(Model2,method="ML"))
```

```
##                                          Model df       AIC       BIC
```

```
## update(Model1, method = "ML")      1 10 412.5837 436.8919
## update(Model2, method = "ML")      2  7 406.5837 423.5994
##                                         logLik    Test      L.Ratio
## update(Model1, method = "ML") -196.2918
## update(Model2, method = "ML") -196.2918 1 vs 2 1.408519e-07
##                                         p-value
## update(Model1, method = "ML")
## update(Model2, method = "ML")             1
```

```
# Model 1 to Model 3
anova(update(Model1,method="ML"), update(Model3,method="ML"))
```

```
##                                    Model df      AIC      BIC
## update(Model1, method = "ML")      1 10 412.5837 436.8919
## update(Model3, method = "ML")      2  7 406.5837 423.5994
##                                         logLik    Test      L.Ratio
## update(Model1, method = "ML") -196.2918
## update(Model3, method = "ML") -196.2918 1 vs 2 1.297511e-06
##                                         p-value
## update(Model1, method = "ML")
## update(Model3, method = "ML")             1
```

```
# Model 2 to Model 4
anova(update(Model2,method="ML"), update(Model4,method="ML"))
```

```
##                                    Model df      AIC      BIC
## update(Model2, method = "ML")      1  7 406.5837 423.5994
## update(Model4, method = "ML")      2  5 402.5837 414.7378
##                                         logLik    Test      L.Ratio
## update(Model2, method = "ML") -196.2918
## update(Model4, method = "ML") -196.2918 1 vs 2 1.239923e-06
##                                         p-value
## update(Model2, method = "ML")
## update(Model4, method = "ML")             1
```

```
# Model 3 to Model 4
anova(update(Model3,method="ML"), update(Model4,method="ML"))
```

```
##                                    Model df      AIC      BIC
## update(Model3, method = "ML")      1  7 406.5837 423.5994
## update(Model4, method = "ML")      2  5 402.5837 414.7378
##                                         logLik    Test      L.Ratio
## update(Model3, method = "ML") -196.2918
## update(Model4, method = "ML") -196.2918 1 vs 2 8.326424e-08
##                                         p-value
## update(Model3, method = "ML")
## update(Model4, method = "ML")             1
```

It can be seen that the 3-random-effects *Model 1* is not statistically significantly different from the reduced 2-random-effects *Model 2* and *Model 3*. Furthermore, the reduced 2-random-effects *Model 2* and *Model 3* are not statistically significantly different from the 1-random-effect *Model 4*. Therefore, *Model 4* is the most parsimonious model for this data. This means that the 3-random-effects model in (7.3) is reduced to one random effect on H_{max} only as follows:

$$\text{Level 2:} \quad H_{max,i} = H_{max} + v_{H_{max,i}}$$

$$r_i = r$$

$$t_{mid,i} = H_{mid}. \tag{7.4}$$

We can print the summary of model fit of *Model 4* as follows:

```
# Print the summary of the best model fitting
summary(Model4)
```

```
## Nonlinear mixed-effects model fit by maximum likelihood
##   Model: height ~ Hmax/(1 + exp(-r * (age - tmid)))
##   Data: Loblolly
##        AIC      BIC    logLik
##   402.5837 414.7378 -196.2918
##
## Random effects:
##  Formula: Hmax ~ 1 | Seed
##              Hmax Residual
## StdDev: 1.950539 2.306948
##
## Fixed effects: Hmax + tmid + r ~ 1
##          Value Std.Error DF  t-value p-value
## Hmax 61.23769 1.0139527 68 60.39501       0
## tmid 11.75006 0.2110242 68 55.68108       0
## r     0.23272 0.0081380 68 28.59675       0
##
## Number of Observations: 84
## Number of Groups: 14
```

A visualization of the model fitting can be seen from the augmented predictions in Fig. 7.3. In this figure, we show the observed data (in dots) with the population prediction (i.e., *fixed* where the random effects are equal to zero) and the within-tree prediction (i.e., *Seed* where the estimates are obtained using the random effects).

```
# plot to confirm it
plot(augPred(Model4, level=0:1))
```

In summary, from this best model fit, we can see that the estimated *within-tree* standard deviation is estimated at $\hat{\sigma} = 2.307$ and the *between*-tree standard deviation $\hat{\tau}_{H_{max}} = 1.951$. With this best fit, the estimated overall fixed-effects parameters are $\hat{H}_{max} = 61.238$, $\hat{r} = 23.272\%$, and $\hat{t}_{mid} = 11.750$. Therefore, with the data from these 14 trees, the overall growth rate is estimated at 23.272% every year, and the maximum height can reach 61.238 (ft). The time for these trees to reach half of the maximum height is estimated at 11.75 (yr).

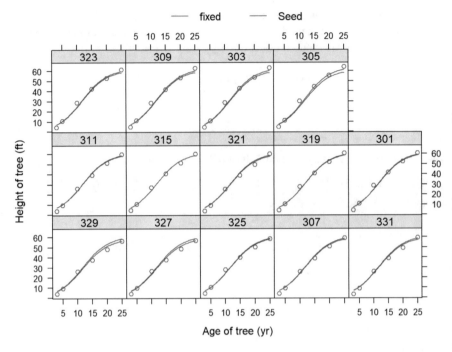

Fig. 7.3 Illustration of model fitting

7.4 Exercise

Use the data *Orange* (growth of orange trees) from *nlme* package to practice what described in this chapter. The data can be accessed:

```
# Access the data
data(Orange)
# Print the data summary
summary(Orange)
```

```
##   Tree         age           circumference
##   3:7    Min.   : 118.0    Min.   : 30.0
##   1:7    1st Qu.: 484.0    1st Qu.: 65.5
##   5:7    Median :1004.0    Median :115.0
##   2:7    Mean   : 922.1    Mean   :115.9
##   4:7    3rd Qu.:1372.0    3rd Qu.:161.5
##          Max.   :1582.0    Max.   :214.0
```

Chapter 8
The Generalized Linear Model

In the previous chapters, we focused on data analysis with a continuous outcome in multi-level data structures, as well as their implementation using the *R* package *nlme*. We now switch to data analysis for non-normal data.

To present multi-level modeling for non-normal data, we will use this chapter to review logistic regression for binary/binomial data and Poisson regression for counts data. These models are part of the *generalized linear model* (*GLM*) which has been widely introduced and well explained, see for example McCullagh and Nelder (1995), Dobson and Barnett (2018), Dunn and Smyth (2018), and Wilson and Chen (2021).

With this review, we will proceed to multi-level logistic regression and multi-level Poisson regression, which are commonly known as generalized linear mixed-effects models (*GLMM*)

Note to readers: We will use *R* package *plot3D* and *MASS* in this chapter. Remember to install these *R* package to your computer using *install.packages(c("plot3D", "MASS"))* and load these packages into *R* session using *library(plot3D)* and *library(MASS)* before running all the *R* programs in this chapter.

8.1 Logistic Regression for Dichotomous Outcome

8.1.1 Data in Medical Diagnostics: The Stress Test

A stress test, sometimes called a treadmill test or exercise test, is typically used by medical doctors to diagnose how well your heart works to handle stress. It is usually done by patients walking on a treadmill or riding a stationary bike with their heart rhythm, blood pressure, and breathing monitored. As their body works harder

© The Author(s), under exclusive license to Springer Nature Switzerland AG 2021
D.-G. (Din) Chen, J. K. Chen, *Statistical Regression Modeling with R*,
Emerging Topics in Statistics and Biostatistics,
https://doi.org/10.1007/978-3-030-67583-7_8

during the test, it requires more oxygen, so the heart must pump more blood. The test can show if the blood supply is reduced in the arteries that supply blood to the heart. This stress test shows how your heart works during extensive physical activity, which can reveal problems with blood flow within your heart.

The doctor may recommend a stress test if you have signs or symptoms of coronary artery disease or irregular heart rhythm (arrhythmia). The test may also guide treatment decisions, measure the effectiveness of treatment, or determine the severity if you have already been diagnosed with a heart condition.

In this chapter, we will use a dataset from Finch et al. (2019) to evaluate whether or not the stress test is effective using logistic regression. In this example, there is a sample of 20 men, 10 of whom have been diagnosed with coronary artery disease and 10 who have not. Each of the 20 individuals was asked to walk on a treadmill until he became too fatigued to continue. The outcome variable in this example was the diagnosis (i.e., *group*) and the independent variable was the time in seconds (i.e., *time*) walked on the treadmill until fatigue. The purpose of this example is to find a model predicting coronary artery status as a function of time walked until fatigue. If an accurate predictive equation could be developed, it might be a helpful tool for physicians to use to help diagnose medical issues from the heart.

Let us look at the data:

```
# read the data into R
dcoronary = read.csv("coronaryArtery.csv", header=T)
# print the data
dcoronary
```

```
##      time group
## 1    1014     0
## 2     684     0
## 3     810     0
## 4     990     0
## 5     840     0
## 6     978     0
## 7    1002     0
## 8    1110     0
## 9     864     1
## 10    636     1
## 11    638     1
## 12    708     1
## 13    786     1
## 14    600     1
## 15    720     1
## 16    750     1
## 17    594     1
## 18    750     1
## 19   1112     0
```

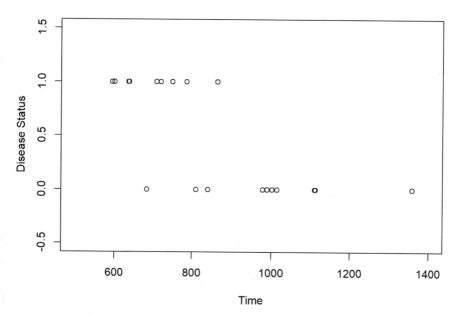

Fig. 8.1 Medical diagnostic data in coronary artery

```
## 20 1359     0
```

As seen from the output, the status (i.e., *group*) of coronary artery disease is coded as *1* for patients with the disease and *0* for patients without the disease.

Let's plot the data using following R code, which shows the Fig. 8.1:

```
plot(group~time, dcoronary, xlim=c(500,1400),
    ylim=c(-0.5,1.5), xlab="Time", ylab="Disease Status")
```

8.1.2 Why Do We Not Model Binary Outcome Data Using Linear Regression?

The simple linear regression can be done as follows:

```
# Fit a linear regression
coronary.lm = lm(group~time, data=dcoronary)
# Print the model fit
summary(coronary.lm)
```

```
##
## Call:
```

```
## lm(formula = group ~ time, data = dcoronary)
##
## Residuals:
##      Min       1Q    Median       3Q       Max
## -0.79458 -0.22617   0.04844  0.28392   0.53022
##
## Coefficients:
##                 Estimate Std. Error t value Pr(>|t|)
## (Intercept)    2.0288200  0.3593292   5.646 2.34e-05 ***
## time          -0.0018044  0.0004129  -4.370 0.000369 ***
## ---
## Signif. codes:  0 '***' 0.001 '**' 0.01 '*' 0.05 '.' 0.1 ' ' 1
##
## Residual standard error: 0.3671 on 18 degrees of freedom
## Multiple R-squared:  0.5148, Adjusted R-squared:  0.4879
## F-statistic:  19.1 on 1 and 18 DF,  p-value: 0.000369
```

Even if we have a significant model with R^2 of 0.51, this model is incorrect and poses an important problem because the predicted values can be outside the range of the expect values of the outcome, 0 and 1 for times fall outside of the 600 and 1400 s as seen in Fig. 8.2. This is important methodological shortfall in using linear regression to model binary data because any values outside of 0 and 1 are not possible.

```
# Plot the data
plot(group~time, dcoronary, xlim=c(500,1400),
    ylim=c(-0.5,1.5), xlab="Time", ylab="Disease Status")
# Add the fitted linear regression line
abline(coronary.lm, col="red", lwd=3)
```

8.1.3   The Logistic Regression Model

To emphasize, any modeling on the dichotomous outcome variable, y (i.e., $y = 0$ or 1) with one or more independent variables (continuous or categorical), the multiple linear regression to the p independent variables:

$$y = \beta_0 + \beta_1 x_1 + \cdots + \beta_p x_p \tag{8.1}$$

is not appropriate anymore due to the binary outcome values of 0 and 1.

To correctly model the binary (or binomial) data, the logistic regression is used to model the probability with logit transformation as follows:

$$log\left(\frac{p}{1-p}\right) = \beta_0 + \beta_1 x_1 + \cdots + \beta_p x_p, \tag{8.2}$$

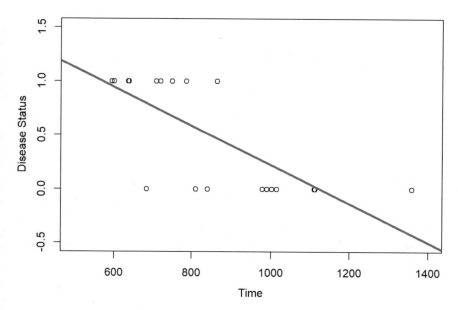

Fig. 8.2 Fitted linear model

where $p = Prob(Y = 1)$, and Y is the outcome variable of interest taking the values 1 or 0 where 1 is typically the outcome of interest. This outcome is linked to independent variables $(x_1, \cdots, x_p)$ by the slope parameter β_i, $i = 0, 1, \cdots, p$ with a logit link function of $log\left(\frac{p}{1-p}\right)$, where $log\left(\frac{p}{1-p}\right)$ is the so-called log odds.

8.1.4   Maximum Likelihood Estimation

In this situation the response variable Y_i for $i = 1, \cdots, n_i$ is binomially distributed (more specifically binary for this data) with n_i fixed, and probability p_i, which is denoted by $Y_i \sim B(n_i, p_i)$. The probability distribution can then be written as

$$P(Y_i = y_i) = \binom{n_i}{y_i} p_i^{y_i} (1 - p_i)^{n_i - y_i}. \tag{8.3}$$

The binomial response variable Y_i may be related to q covariates, such as treatment received, sex, age, etc., which are denoted by $(x_{i1}, \cdots, x_{iq})$. The fundamental difference between logistic regression and multiple linear regression is that the response variable (i.e., outcome variable) is binomially distributed, and it is not normally distributed anymore. We then model the probability p_i as a linear function of the q covariates. This is the basic idea behind logistic regression and is described as part of the *generalized linear model* developed for the exponential family of distributions.

In the generalized linear model framework, we use *a linear predictor* to model the linear relationship and a *link function* to link the *linear predictor* to the binomial probability p_i. Specifically, the *linear predictor* is denoted by

$$\eta_i = \beta_0 + \beta_1 x_{i1} + \cdots + \beta_q x_{iq} = X_i \beta, \tag{8.4}$$

where $X_i = (1, x_{i1}, \cdots, x_{iq})$ is the matrix of observed covariates and $\beta = (\beta_0, \beta_1, \cdots, \beta_q)$ is the associated parameter vector.

Various *link functions* are possible for linking the *linear predictor* η, which is a linear combination of the effects of one or more explanatory variables, to the (outcome) probabilities p_i that we want to model. It is easy to see that the identity link of $p_i = \eta_i$ is not appropriate since the binomial probability p_i has to be constrained to the $[0, 1]$ interval. For binomial response, the most commonly used link function is the so-called *logit* (and therefore giving rise to the term logistic regression) given by

$$\eta = log \left(\frac{p}{1-p} \right) \tag{8.5}$$

or

$$p = \frac{e^\eta}{1 + e^\eta}. \tag{8.6}$$

Other link functions used in the logistic regression of binomial response data are the *probit link function*: $\eta = \Phi^{-1}(p)$, where Φ^{-1} is the inverse normal cumulative distribution function, and the *complementary log-log link function*: $\eta = log[-log(1 - p)]$.

Maximum likelihood methods may be used for parameter estimation and statistical inference. We briefly describe maximum likelihood estimation (MLE) as follows.

To perform MLE, we first specify the likelihood function to be maximized. The general likelihood function is defined as

$$P(Y = y) = \prod_{i=1}^{n} f(y_i | \theta) = L(\theta | y). \tag{8.7}$$

For binomial data, the likelihood becomes

$$L(\beta | y) = \prod_{i=1}^{n} \binom{n_i}{y_i} p_i^{y_i} (1 - p_i)^{n_i - y_i}. \tag{8.8}$$

The likelihood is a function of the unknown parameters, the observed response data, and the covariates. MLE of the unknown parameters requires finding values of the parameters that maximize the likelihood function.

Maximizing the likelihood function is equivalent to maximizing the log-likelihood function:

$$l(\theta|y) = log L(\theta|y). \tag{8.9}$$

For binomial data:

$$l(\beta|y) = \sum_{i=1}^{n} \left[log \binom{n_i}{y_i} + y_i log(p_i) + (n_i - y_i) log(1 - p_i) \right]. \tag{8.10}$$

If we use the *logit link function* of $\eta = log\left(\frac{p}{1-p}\right)$, then $p = \frac{e^{\eta}}{1+e^{\eta}}$. The log-likelihood function becomes

$$
\begin{aligned}
l(\beta|y) &= \sum_{i=1}^{n} \left[log \binom{n_i}{y_i} + y_i log(p_i) + (n_i - y_i) log(1 - p_i) \right] \\
&= \sum_{i=1}^{n} \left[y_i \eta_i - n_i log(1 + e^{\eta_i}) + log \binom{n_i}{y_i} \right].
\end{aligned} \tag{8.11}
$$

This function is then maximized to obtain the parameter estimates. In software implementation in optimization, minimization is typically implemented and therefore the theory of maximum likelihood estimation to maximize the log-likelihood function is in fact to minimize the negative log-likelihood function.

There is no analytical closed-form solution for the parameter estimates as there was in normal based multiple linear regression in Chap. 1. Numerical search methods are required and maximum likelihood estimation theory is drawn upon for parameter estimates and their standard errors, confidence intervals, *p*-values as well as model selection. This logistic regression is implemented in *R* as a function *glm*.

8.1.5 Illustration of the Likelihood Function

To understand the maximum likelihood estimation, it is essential to understand the multi-dimensional surface of the likelihood function corresponding to the different values of the parameters. To demonstrate this property, let us use this *coronary* data as an example since there are only two parameters (i.e., intercept β_0 and slope β_1) in this simple logistic regression. By changing these two parameters, the negative log-likelihood function would be displayed as a 3-dimensional surface.

```
# Make the negative log-likelihood function
negll = function(b0, b1){
  # Get the data
  x = dcoronary$time; y = dcoronary$group
  # Calculate the probrobility p
  eta = b0+b1*x; p = exp(eta)/(1+exp(eta))
  # The log-likelihood function
  ll = y*log(p)+(1-y)*log(1-p);
  -sum(ll)
}

# Specify parameter sequences
npts = 10 # number of points
b0=seq(12,14,length=npts);b1=seq(-0.02,-0.012,length=npts)
# Loop over to calculate the negative log-likelihood
negll.val = matrix(0, ncol=npts, nrow=npts)
for(i in 1:npts){
  for (j in 1:npts){
    negll.val[i,j] = negll(b0[i],b1[j]) }}
colnames(negll.val)= round(b1,3);
rownames(negll.val)= round(b0,2);
# Print the values
round(negll.val,2)
```

```
##        -0.02 -0.019 -0.018 -0.017 -0.016 -0.016 -0.015 -0.014
## 12     23.67  18.88  14.70  11.24   8.65   7.05   6.52   7.12
## 12.22  21.96  17.39  13.45  10.28   8.02   6.75   6.57   7.52
## 12.44  20.33  15.97  12.30   9.43   7.49   6.57   6.73   8.02
## 12.67  18.77  14.65  11.25   8.69   7.08   6.49   6.99   8.61
## 12.89  17.30  13.43  10.31   8.06   6.77   6.52   7.35   9.30
## 13.11  15.92  12.30   9.47   7.53   6.58   6.66   7.80  10.09
## 13.33  14.62  11.27   8.73   7.12   6.49   6.89   8.35  10.96
## 13.56  13.42  10.34   8.11   6.81   6.50   7.21   8.98  11.93
## 13.78  12.31   9.51   7.59   6.61   6.61   7.63   9.71  12.98
## 14     11.30   8.79   7.17   6.50   6.81   8.13  10.52  14.12
##        -0.013 -0.012
## 12       8.93  12.07
## 12.22    9.68  13.18
## 12.44   10.53  14.39
## 12.67   11.48  15.69
## 12.89   12.51  17.06
## 13.11   13.64  18.52
## 13.33   14.86  20.05
## 13.56   16.16  21.65
## 13.78   17.54  23.32
## 14      18.99  25.03
```

```
# Let's find out which beta0 and beta1 to make the minimum
min.val = which.min(negll.val)
est.beta0 = b0[min.val%%npts]; est.beta0
```

```
## [1] 13.33333
```

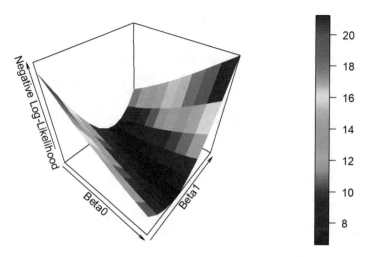

Fig. 8.3 Surface for the negative log-likelihood function

```
est.beta1 = b1[min.val%/%npts+1]; est.beta1
```

```
## [1] -0.01644444
```

```
# What's the minimum
min.negll = negll.val[min.val%%npts,min.val%/%npts+1]
min.negll
```

```
## [1] 6.490215
```

Notice in the above *R* code chunk, we identify the minimum value of the negative log-likelihood function from all the (β_0, β_1) combinations using *min.val = which.min(negll.val)*), which is *6.4902*. Corresponding to this minimum, we find the corresponding (β_0, β_1) value using *est.beta0 = b0[min.val%%npts]* and *est.beta1 = b1[min.val%/%npts+1]*, which is $(\beta_0, \beta_1) = (13.3333, -0.01644)$. This is the basic steps to get the maximum likelihood estimation in numerical optimization.

The graphical illustration can be produced in Fig. 8.3 as follows:

```
# Load the package for 3D plotting
library(plot3D)

# Plot the 3D surface
persp3D(b0,b1,negll.val,
        xlab="Beta0",ylab="Beta1",
        zlab="Negative Log-Likelihood")
```

8.1.6   R *Implementation in* glm

Logistic regression models can be fit easily in *R* using the *glm* as follows:

```
# fit logistic regression
coronary.logistic<-glm(group~time, data=dcoronary, family=binomial)
# print the fit for logistic regression
summary(coronary.logistic)
```

```
##
## Call:
## glm(formula = group ~ time, family = binomial, data = dcoronary)
##
## Deviance Residuals:
##     Min       1Q   Median       3Q      Max
## -2.1387  -0.3077   0.1043   0.5708   1.5287
##
## Coefficients:
##               Estimate Std. Error z value Pr(>|z|)
## (Intercept) 13.488949   5.876693   2.295   0.0217 *
## time        -0.016534   0.007358  -2.247   0.0246 *
## ---
## Signif. codes:  0 '***' 0.001 '**' 0.01 '*' 0.05 '.' 0.1 ' ' 1
##
## (Dispersion parameter for binomial family taken to be 1)
##
##     Null deviance: 27.726  on 19  degrees of freedom
## Residual deviance: 12.966  on 18  degrees of freedom
## AIC: 16.966
##
## Number of Fisher Scoring iterations: 6
```

In logistic regression, we examine the deviance value to assess the model fit to the data. Deviance is a measure of goodness of fit of a generalized linear model. Or rather, it is a measure of *badness* of fit, with higher values indicating a worse or inaccurate fit to the data.

As seen from the output, *R* reports two types of deviance—the null deviance and the residual deviance. The null deviance shows how well the response variable (i.e., *group* in this example) is predicted by a model that includes only the intercept (grand mean). In this example, we have a value of 27.726 on 19 degrees of freedom. Including the independent variables (i.e., *time*) decreased the deviance to 12.966 on 18 degrees of freedom, a significant reduction in deviance confirming that *time* is a significant predictor of coronary status.

Since the residual deviance is χ^2 distributed, we can calculate the *p*-value associated with this distribution. As seen from this logistic regression, the deviance from the model fit is 12.966 with 18 degrees of freedom, we can extract these values and calculate the *p*-value as follows:

```
# Extract the deviance
dev.logistic = deviance(coronary.logistic);
dev.logistic
```

```
## [1] 12.96558
```

```
# Extract the df
df.logistic  = coronary.logistic$df.residual;
df.logistic
```

```
## [1] 18
```

```
# Calculate the p-vlaue
pval = 1-pchisq(dev.logistic, df.logistic)
pval
```

```
## [1] 0.7936135
```

As seen from the calculation, the p-value = 0.7936 indicates that this logistic regression model fit the data very well. The estimated intercept is $\hat{\beta}_0 = 13.4889$ and the estimated slope $\hat{\beta}_1$ is -0.016534, which is very close to the illustration in Sect. 8.1.5 where $(\beta_0, \beta_1) = (13.3333, -0.01644)$ was identified from the figure.

Furthermore, the ith p-value associated with $\hat{\beta} = 0.0246$, thus indicating that a significant relationship between the independent variable (*time*) and the dependent (coronary disease status) exists. Specifically, this means that the more time a patient could walk on the treadmill before becoming fatigued, the less likely they would have had heart disease.

Attention should be paid to the explanation of estimated parameters. In logistic regression, the intercept parameter β_0 is the log odds when the independent variable (i.e., *time* in this data) is zero. The slope parameter β_1 is the log odds-ratio when the independent variable increases by one unit. To get the estimated odds, an anti-log (i.e., exponential) transformation should be used for these parameters to get the estimated odds. For example, taking the anti-log to the estimated slope $e^{\hat{\beta}_1}$, we can obtain the odds of having coronary artery disease as a function of time. In this example, the $\hat{\beta}_1 = -0.016534$, which is negative. This means that the more time an individual can walk, the lower the log odds is that (s)he had coronary artery disease. Since $e^{-0.016534} = 0.984$, this means that for every additional second an individual can walk on a treadmill before becoming fatigued, her/his estimated odds of having heart disease are multiplied by 0.984 (i.e., decreased). Thus, for an additional minute of walking, the odds of having coronary artery disease decrease by $exp(-0.016534 \times 60) = 0.378$.

With this fitted logistic regression, we can add the fit to the data to illustrate the model fitting in Fig. 8.4 as follows:

```
# plot the data
plot(group~time, dcoronary, xlim=c(500,1400),
    ylim=c(0,1), xlab="Time", ylab="Coronary Disease Status")

# make a temp time sequence
tm = seq(500, 1400, length=1000)
# get the estimated parms
est.parm = coef(coronary.logistic)
# calculate the linear part
right.side = est.parm[1]+est.parm[2]*tm
# transform to probs
p = exp(right.side)/(1+exp(right.side))

# add all those points as line to the plot
lines(tm, p,col="red", lwd=3)
```

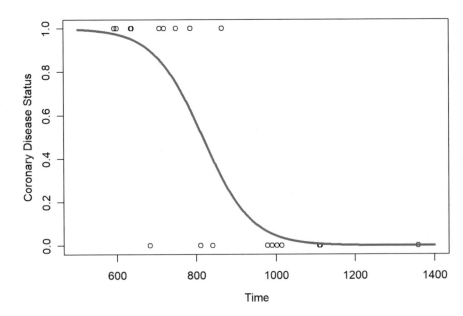

Fig. 8.4 Fitted logistic regression

8.2 Poisson Regression: Models for Counts Data

8.2.1 Research Papers by Graduate Students in Biochemistry PhD Programs

To illustrate the regression for counts data, let us consider a dataset typically used for Poisson regression. This dataset is about the number of research articles published by graduate students in biochemistry PhD programs. This data can be accessed from *R* library *pscl* (i.e., Political Science Computational Laboratory from Stanford University) as follows:

```r
# load the library
library(pscl)
# load the data
data(bioChemists)
# dimmension
dim(bioChemists)
```

```
## [1] 915    6
```

```r
# summary of the data
summary(bioChemists)
```

```
##       art              fem            mar           kid5
##  Min.   : 0.000   Men  :494   Single :309   Min.   :0.0000
##  1st Qu.: 0.000   Women:421   Married:606   1st Qu.:0.0000
##  Median : 1.000                             Median :0.0000
##  Mean   : 1.693                             Mean   :0.4951
##  3rd Qu.: 2.000                             3rd Qu.:1.0000
##  Max.   :19.000                             Max.   :3.0000
##       phd            ment
##  Min.   :0.755   Min.   : 0.000
##  1st Qu.:2.260   1st Qu.: 3.000
##  Median :3.150   Median : 6.000
##  Mean   :3.103   Mean   : 8.767
##  3rd Qu.:3.920   3rd Qu.:12.000
##  Max.   :4.620   Max.   :77.000
```

The variables for this dataframe are

- *art* = count of articles produced during last 3 years of PhD
- *fem* = factor indicating gender of student, with levels Men and Women
- *mar* = factor indicating marital status of student, with levels Single and Married
- *kid5* = number of children aged 5 or younger
- *phd* = prestige of PhD department
- *ment* = count of articles produced by PhD mentor during last 3 years

The purpose of using this data is to examine which factors (i.e., *fem, mar, kid5, phd*, and *ment*) are significant predictors to the outcome *art*.

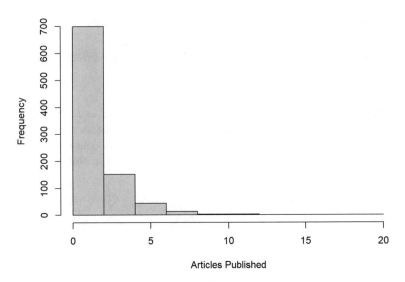

Fig. 8.5 Histogram of the number of articles published

To view the distribution of the number of articles produced during last 3 years of each individual's PhD (i.e., *art*), we can use the *hist(art)*:

```
hist(bioChemists$art, xlab="Articles Published", main="")
```

As seen in Fig. 8.5, the distribution of the number of articles published is not normally distributed, and it is highly skewed to the right. This is not a surprise since counts data are typically non-negative and typically skewed.

8.2.2   Again Why Not Classical Linear Regression

As usual, we fit a linear regression:

```
# Fit the linear regression
art.lm =lm(art~., bioChemists)
# Print the model fit
summary(art.lm)
```

```
##
## Call:
## lm(formula = art ~ ., data = bioChemists)
##
## Residuals:
##     Min      1Q  Median      3Q     Max
## -5.0209 -1.2358 -0.4125  0.7517 14.8409
##
```

```
## Coefficients:
##              Estimate Std. Error t value Pr(>|t|)
## (Intercept)  1.334249   0.240911   5.538    4e-08 ***
## femWomen    -0.380885   0.128139  -2.972  0.00303 **
## marMarried   0.263199   0.145423   1.810  0.07064 .
## kid5        -0.291442   0.090919  -3.206  0.00140 **
## phd         -0.011365   0.063666  -0.179  0.85836
## ment         0.061495   0.006605   9.310  < 2e-16 ***
## ---
## Signif. codes:  0 '***' 0.001 '**' 0.01 '*' 0.05 '.' 0.1 ' ' 1
##
## Residual standard error: 1.822 on 909 degrees of freedom
## Multiple R-squared:  0.1104, Adjusted R-squared:  0.1055
## F-statistic: 22.55 on 5 and 909 DF,  p-value: < 2.2e-16
```

We can check the residual distribution using QQ-plot as shown in Fig. 8.6 as follows:

```
qqnorm(art.lm$residuals)
qqline(art.lm$residuals, lwd=3, col="red")
```

We can see a large deviation from the normal distribution in Fig. 8.6, which is indeed the case since the data are counts. Therefore, the classical linear regression

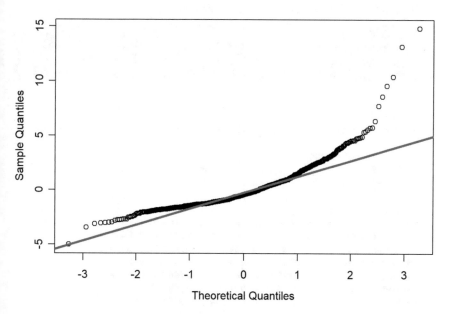

Fig. 8.6 QQ-plot of the residual from linear regression

is not appropriate to model counts data and a new regression model should be developed for counts data. This is the Poisson regression.

8.2.3 Poisson Regression

As demonstrated in the above example on the counts of articles published by graduate students, such a variable is not normally distributed anymore since they are counts and must be non-negative. This presents problems in applying the standard linear regression model, since classical linear regression may produce predicted counts that are less than 0 and have residuals that are very deviated from the assumed normal distribution. This can be seen in Fig. 8.6.

To deal with these potential problems, the Poisson regression was developed as part of the generalized linear model (GLM) with a log link function to overcome the problem of negative predicted counts, since the log of the counts can take any real numbered value as follows:

$$ln(Y) = \beta_0 + \beta_1 x_1 + \cdots + \beta_p x_p, \tag{8.12}$$

where Y is the counts outcome and $x_1, \cdots, x_p$ are the p predictors (i.e., $x_1 = fem$, $x_2 = mar$, $x_3 = kid5$, $x_4 = phd$, and $x_5 = ment$ in this example). $\beta_0, \beta_1, \cdots, \beta_p$ are the parameters to be estimated.

We can fit the Poisson regression model as follows:

```
# Fit the Poisson regression
art.poisson = glm(art~., data=bioChemists, family =c("poisson"))
# Print the model fit
summary(art.poisson)
```

```
##
## Call:
## glm(formula = art ~ ., family = c("poisson"), data = bioChemists)
##
## Deviance Residuals:
##     Min       1Q   Median       3Q      Max
## -3.5672  -1.5398  -0.3660   0.5722   5.4467
##
## Coefficients:
##               Estimate Std. Error z value Pr(>|z|)
## (Intercept)   0.304617   0.102981   2.958   0.0031 **
## femWomen     -0.224594   0.054613  -4.112 3.92e-05 ***
## marMarried    0.155243   0.061374   2.529   0.0114 *
## kid5         -0.184883   0.040127  -4.607 4.08e-06 ***
## phd           0.012823   0.026397   0.486   0.6271
## ment          0.025543   0.002006  12.733  < 2e-16 ***
## ---
## Signif. codes:  0 '***' 0.001 '**' 0.01 '*' 0.05 '.' 0.1 ' ' 1
##
```

```
## (Dispersion parameter for poisson family taken to be 1)
##
##       Null deviance: 1817.4  on 914   degrees of freedom
## Residual deviance: 1634.4  on 909   degrees of freedom
## AIC: 3314.1
##
## Number of Fisher Scoring iterations: 5
```

As seen from the above *R* code chunk, we used probability distribution *family* =c("*poisson*") in *glm* to fit the Poisson regression. The result showed that the gender status *female*, married status *mar*, kids status *kids*, and mentor's publications *ment* are significant. However, before we make any conclusion, we need to examine whether or not this model fits the data by examining the residual deviance. As discussed in the logistic regression, the residual deviation is χ^2 distributed and can be used to test the null hypothesis that the model fits the data. We extract the residual deviance and the associated degrees of freedom for this calculation as follows:

```
## p-value
pval.poisson= 1-pchisq(deviance(art.poisson),
               df.residual(art.poisson))
# print
pval.poisson
```

```
## [1]  0
```

The resulting *p*-value < 0.0001, which is highly significant, suggesting that the Poisson model *art.poisson* did not fit the data adequately and that further exploration is needed. This lack of goodness of fit is probably due to the data collection where there are other important predictive variables missing and they are correlated with the number of articles published by the students. For the purpose of modeling demonstration, let us use this data for further model fitting.

8.2.4   Models for Overdispersed Counts Data

The fundamental assumption in Poisson regression is that the counts are assumed to be Poisson distributed and the underlying assumption of Poisson distribution is that the mean is equal to the variance, i.e., $\sigma^2 = var(Y) = mean(Y) = \mu$.

To verify whether this is the case for this data, we can calculate them as follows:

```
# Calculate the variance and mean
var.art = var(bioChemists$art); var.art
```

```
## [1]  3.709742
```

```
mean.art = mean(bioChemists$art); mean.art
```

```
## [1] 1.692896
```

```
# the ratio between the variance and mean
var.art/mean.art
```

```
## [1] 2.191358
```

From this calculation, the variance is 3.7097 and the mean is 1.6929. The variance is 2.2 times larger than the mean, indicating that the fundamental assumption of Poisson distribution is violated. This phenomenon is called overdispersion in counts regression.

To deal with overdispersion, two approaches are usually used with one called *Quasi-Poisson regression* and another called *negative-binomial regression*.

8.2.4.1 Quasi-Poisson Regression

In quasi-Poisson regression, the overdispersion is incorporated with the overdispersion parameter ϕ, i.e., $\sigma^2 = var(Y) = \phi \times mean(Y) = \phi\mu$. Therefore when ϕ is greater than 1, we can model the over-dispersion and when ϕ is less than 1, we can model the under-dispersion (another phenomenon in counts data). The classical Poisson regression is then the special case that $\phi = 1$. Therefore, the quasi-Poisson regression is an extension of the classical Poisson regression. This overdispersion parameter will be estimated along with all the βs from the maximum likelihood estimation.

In fitting the quasi-Poisson regression, the overdispersion parameter is estimated based on the data and then used to adjust the standard errors for all the estimated parameters. In this direction, the estimated values of the parameters βs will not change. The quasi-Poisson regression can be implemented in *glm* as follows:

```
art.quasipoisson = glm(art~., data=bioChemists, family=c("quasipoisson"))
# summary
summary(art.quasipoisson)
```

```
##
## Call:
## glm(formula = art ~ ., family = c("quasipoisson"), data = bioChemists)
##
## Deviance Residuals:
##     Min       1Q   Median       3Q      Max
## -3.5672  -1.5398  -0.3660   0.5722   5.4467
##
## Coefficients:
##               Estimate Std. Error t value Pr(>|t|)
## (Intercept)   0.304617   0.139273   2.187 0.028983 *
## femWomen     -0.224594   0.073860  -3.041 0.002427 **
## marMarried    0.155243   0.083003   1.870 0.061759 .
```

```
## kid5        -0.184883   0.054268  -3.407 0.000686 ***
## phd          0.012823   0.035700   0.359 0.719544
## ment         0.025543   0.002713   9.415  < 2e-16 ***
## ---
## Signif. codes:  0 '***' 0.001 '**' 0.01 '*' 0.05 '.' 0.1 ' ' 1
##
## (Dispersion parameter for quasipoisson family taken to be 1.829006)
##
##     Null deviance: 1817.4  on 914  degrees of freedom
## Residual deviance: 1634.4  on 909  degrees of freedom
## AIC: NA
##
## Number of Fisher Scoring iterations: 5
```

As seen from the output, the dispersion parameter is estimated at 1.829 as seen *(Dispersion parameter for quasi-Poisson family taken to be 1.829006)* in the *art.quasipossion* model where the dispersion parameter is defaulted to 1 in the classical Poisson regression *art.poisson* as seen *(Dispersion parameter for Poisson family taken to be 1)*. With the estimated dispersion parameter of 1.829, we can conclude that there is an overdispersion in the counts of articles published by graduate students. Adjusting this estimated dispersion parameter, the standard errors of all the parameters are scaled by 1.829, which is larger than those in the classical Poisson regression *art.poisson*. With this adjustment, the statistical significance for all parameters are reduced. Noticeably the *p*-value for marriage status *mar* changed from 0.0114 (statistically significant) to 0.0617 (statistically not significant).

8.2.4.2 Negative-Binomial Regression

Another alternative to the classical Poisson regression when data are overdispersed is the negative-binomial regression model based on the negative-binomial distribution. In the negative-binomial distribution, the mean is identical to that of the Poisson; the variance is

$$Var(Y) = \mu + \frac{\mu^2}{\theta}. \tag{8.13}$$

As seen from the above equation, an extra term $\frac{\mu^2}{\theta}$ is added to the mean μ. In this model, it is clear that as θ increases and approaches infinity, the variance approaches the mean. In this case, the negative-binomial distribution becomes more like the classical Poisson distribution. When $\theta = 1$, this negative-binomial distribution becomes the Gamma distribution. In this special case, the negative-binomial regression will be the Gamma regression.

The negative-binomial regression is implemented in *R* function *glm.nb* within the *MASS* library (i.e., functions and datasets to support Venables and Ripley, *Modern Applied Statistics with S*). For this example, the implementation to fit the negative-binomial regression is as follows:

```
# Load the MASS library
library(MASS)
# fit negative binomial model
art.nb = glm.nb(art~., data=bioChemists)
# summary
summary(art.nb)
```

```
##
## Call:
## glm.nb(formula = art ~ ., data = bioChemists, init.theta = 2.264387695,
##     link = log)
##
## Deviance Residuals:
##     Min       1Q   Median       3Q      Max
## -2.1678  -1.3617  -0.2806   0.4476   3.4524
##
## Coefficients:
##             Estimate Std. Error z value Pr(>|z|)
## (Intercept)  0.256144   0.137348   1.865 0.062191 .
## femWomen    -0.216418   0.072636  -2.979 0.002887 **
## marMarried   0.150489   0.082097   1.833 0.066791 .
## kid5        -0.176415   0.052813  -3.340 0.000837 ***
## phd          0.015271   0.035873   0.426 0.670326
## ment         0.029082   0.003214   9.048  < 2e-16 ***
## ---
## Signif. codes:  0 '***' 0.001 '**' 0.01 '*' 0.05 '.' 0.1 ' ' 1
##
## (Dispersion parameter for Negative Binomial(2.2644) family taken to be 1)
##
##     Null deviance: 1109.0  on 914  degrees of freedom
## Residual deviance: 1004.3  on 909  degrees of freedom
## AIC: 3135.9
##
## Number of Fisher Scoring iterations: 1
##
##
##               Theta:   2.264
##           Std. Err.:   0.271
##
##  2 x log-likelihood:  -3121.917
```

As shown above, the estimated overdispersion parameter is $\hat{\theta} = 2.264$ with a standard error of 0.271. Similar statistical significance can be observed in the negative-binomial regression to the quasi-Poisson regression.

To examine whether this negative-binomial regression fits the data, we can similarly extract the residual deviance to calculate the p-value from the χ^2-test as follows:

```
## p-value
pval.nb = 1-pchisq(deviance(art.nb),
            df.residual(art.nb))
# print
pval.nb
```

```
## [1] 0.01475755
```

This small p-value indicates that this model did not fit the data satisfactorily. This could be due to other unmeasured predictors that were missed during data collection. As a result, further exploration should be carried out.

8.3 Exercises

1. Data *Arrests* from *R* package *effects* is about *Arrests for Marijuana Possession*, which is on police treatment of individuals arrested in Toronto for simple possession of small quantities of marijuana. The data are part of a larger dataset featured in a series of articles in the Toronto Star newspaper. Analyze this data using logistic regression to identify whether *released* (i.e., Whether the arrestee was released with a summon; a factor with levels: No; Yes) is a statistically related variable:

- *color*: The arrestee's race; a factor with levels: Black; White.
- *year*: 1997 through 2002; a numeric vector.
- *age*: in years; a numeric vector.
- *sex*: a factor with levels: Female; Male.
- *employed*: a factor with levels: No; Yes.
- *citizen*: a factor with levels: No; Yes.

2. Analyze the *Arthritis* treatment data from *R* package *vcd* with logistic regression to identify whether *Treatment* and *Sex*, *Age* are statistically related with the treatment outcome.

- *Note*: to dichotomize the outcome *Improved* into 1 = *Better* (i.e., combine *Some* and *Marked* into 1) and 0= *None*.
- *Hint*: load the data using R code: *data("Arthritis", package="vcd")*.

Chapter 9
Generalized Multi-Level Model for Dichotomous Outcome

Having reviewed logistic regression in Chap. 8, we now proceed to the generalized linear mixed-effects model (GLMM) to analyze binary data with multi-level data structures.

In this chapter, we will use the data from Appendix A.9 to illustrate linear logistic regression and multi-level generalized linear regression. Multi-level generalized linear regression is one of the models among the generalized linear mixed-effects models (GLMMs).

Note to readers: We will use *R* package *lme4* in this chapter. Remember to install this *R* package to your computer using *install.packages("lme4")* and load this package into *R* session using *library(lme4)* before running all the *R* programs in this chapter.

9.1 Data on the 1988 USA Presidential Election

As discussed in Appendix A.9, this dataset has 13,544 observations from 51 states (including the *District of Columbia*). There are 8 surveys in this dataset. The last survey (i.e., *survey* = 9158 in the dataset) was conducted in September 1988 and was analyzed by Gelman and Hill (2007) to better predict the election results, since this was the survey data that was collected closest to the election.

In this chapter, we will use the first survey (i.e., *survey* = *9152*) as an example to illustrate logistic regression (discussed in Chap. 8) and GLMM. We will focus on the modeling of several predictors to the probability of the respondent preference to the Republican candidate (i.e., Bush) for president. These predictors include their education level (i.e., *edu*), age category (i.e., *age*), gender (i.e., *female*), and race (i.e., *black*) measured in 50 states and the District of Columbia.

The dataset can be loaded into *R* as follows:

© The Author(s), under exclusive license to Springer Nature Switzerland AG 2021
D.-G. (Din) Chen, J. K. Chen, *Statistical Regression Modeling with R*,
Emerging Topics in Statistics and Biostatistics,
https://doi.org/10.1007/978-3-030-67583-7_9

```
# Read in the polls data#
dPoll = read.csv("polls.csv", header=T)
# Check the dimension of the data
dim(dPoll)
```

```
## [1]  13544      10
```

To analyze data from the first survey, we can subset the full dataset *dPoll* for its first survey as follows:

```
# Subset the data to get the first survey
dSurvey1 = dPoll[dPoll$survey== 9152,]

# Make the outcome, *bush*,  as categorical
dSurvey1$bush = as.factor(dSurvey1$bush)

# Make the "state" as categorical
dSurvey1$state = as.factor(dSurvey1$state)

# Make the "edu" as categorical
dSurvey1$edu = as.factor(dSurvey1$edu)

# Make the "age" as categorical
dSurvey1$age = as.factor(dSurvey1$age)

# Make the gender "female" as categorical
dSurvey1$female = as.factor(dSurvey1$female)

# Make the race "black" as categorical
dSurvey1$black = as.factor(dSurvey1$black)

# keep the variables for this analysis
dSurvey1 = dSurvey1[,
c("bush","state","edu","age","female","black")]

# Check the dimension of the data
dim(dSurvey1)
```

```
## [1]  1611      6
```

```
# Print the summary of the first survey
summary(dSurvey1)
```

```
##    bush          state        edu      age     female   black
##  0  :650     5    : 190    1:191    1:377    0:699    0:1496
##  1  :810     33   :  86    2:573    2:559    1:912    1: 115
```

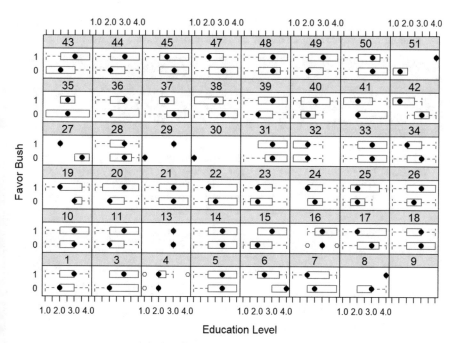

Fig. 9.1 Boxplot in favor of bush by education level

```
##   NA's:151     14        :  82    3:411     3:432
##                44        :  80    4:436     4:243
##                10        :  79
##                36        :  78
##                (Other):1016
```

To get a sense of the *state*-level variability, we can use the *R* package *lattice* to examine relationships within nested structures. For example, to examine the relationship for respondents who are in favor of Bush with their education level, we can call the *bwplot* to plot the distribution as follows:

```r
# load the library
library(lattice)

# Plot the data
bwplot(bush~edu|state,data = dSurvey1, xlab="Education Level", ylab="Favor Bush")
```

As seen in Fig. 9.1, with increased education level, there are more respondents who favor Bush (i.e., the mean *median* value in the boxplot corresponding to *Favor Bush = 1* in each state) in states with ID of 1, 3, 8, 11, 15, 20, 26, 29, 36, 39, 40, 43, 44, 49, 51, but less in the rest of the states.

To know which states correspond to these state IDs, we need to match the IDs to the state names. We can make use of the built-in dataset *state* from *R* as follows:

```
# Load the *state* data
data(state)

# Get their 2-letter abbreviations, add DC as the 9th ordered state
state.abbr = c(state.abb[1:8], "DC", state.abb[9:50])
state.abbr
```

```
##  [1] "AL" "AK" "AZ" "AR" "CA" "CO" "CT" "DE" "DC" "FL" "GA"
## [12] "HI" "ID" "IL" "IN" "IA" "KS" "KY" "LA" "ME" "MD" "MA"
## [23] "MI" "MN" "MS" "MO" "MT" "NE" "NV" "NH" "NJ" "NM" "NY"
## [34] "NC" "ND" "OH" "OK" "OR" "PA" "RI" "SC" "SD" "TN" "TX"
## [45] "UT" "VT" "VA" "WA" "WV" "WI" "WY"
```

```
# Get the state name
state.name = c(state.name[1:8], "District of Columbia", state.name[9:50])
state.name
```

```
##  [1] "Alabama"              "Alaska"
##  [3] "Arizona"              "Arkansas"
##  [5] "California"           "Colorado"
##  [7] "Connecticut"          "Delaware"
##  [9] "District of Columbia" "Florida"
## [11] "Georgia"              "Hawaii"
## [13] "Idaho"                "Illinois"
## [15] "Indiana"              "Iowa"
## [17] "Kansas"               "Kentucky"
## [19] "Louisiana"            "Maine"
## [21] "Maryland"             "Massachusetts"
## [23] "Michigan"             "Minnesota"
## [25] "Mississippi"          "Missouri"
## [27] "Montana"              "Nebraska"
## [29] "Nevada"               "New Hampshire"
## [31] "New Jersey"           "New Mexico"
## [33] "New York"             "North Carolina"
## [35] "North Dakota"         "Ohio"
## [37] "Oklahoma"             "Oregon"
## [39] "Pennsylvania"         "Rhode Island"
## [41] "South Carolina"       "South Dakota"
## [43] "Tennessee"            "Texas"
## [45] "Utah"                 "Vermont"
## [47] "Virginia"             "Washington"
## [49] "West Virginia"        "Wisconsin"
## [51] "Wyoming"
```

```
# make a new data
dState = data.frame(state=1:51, state.abbr = state.abbr, state.name=state.name)

# Print the new data
dState
```

```
##      state state.abbr            state.name
## 1        1         AL               Alabama
## 2        2         AK                Alaska
## 3        3         AZ               Arizona
## 4        4         AR              Arkansas
## 5        5         CA            California
## 6        6         CO              Colorado
## 7        7         CT           Connecticut
## 8        8         DE              Delaware
## 9        9         DC District of Columbia
## 10      10         FL               Florida
## 11      11         GA               Georgia
## 12      12         HI                Hawaii
## 13      13         ID                 Idaho
## 14      14         IL              Illinois
## 15      15         IN               Indiana
## 16      16         IA                  Iowa
## 17      17         KS                Kansas
## 18      18         KY              Kentucky
## 19      19         LA             Louisiana
## 20      20         ME                 Maine
## 21      21         MD              Maryland
## 22      22         MA         Massachusetts
## 23      23         MI              Michigan
## 24      24         MN             Minnesota
## 25      25         MS           Mississippi
## 26      26         MO              Missouri
## 27      27         MT               Montana
## 28      28         NE              Nebraska
## 29      29         NV                Nevada
## 30      30         NH         New Hampshire
## 31      31         NJ            New Jersey
## 32      32         NM            New Mexico
## 33      33         NY              New York
## 34      34         NC        North Carolina
## 35      35         ND          North Dakota
## 36      36         OH                  Ohio
## 37      37         OK              Oklahoma
## 38      38         OR                Oregon
```

```
## 39       39        PA         Pennsylvania
## 40       40        RI         Rhode Island
## 41       41        SC         South Carolina
## 42       42        SD         South Dakota
## 43       43        TN         Tennessee
## 44       44        TX         Texas
## 45       45        UT         Utah
## 46       46        VT         Vermont
## 47       47        VA         Virginia
## 48       48        WA         Washington
## 49       49        WV         West Virginia
## 50       50        WI         Wisconsin
## 51       51        WY         Wyoming
```

With the new data *dState*, we can see the states' name for states in favor of Bush based on the distribution in Fig. 9.1:

```
# State's ID in favor of Bush
states4Bush = c(1, 3, 8, 11, 15, 20, 26, 29, 36, 39,
40, 43, 44, 49, 51)

# States' name in favor of Bush
dState$state.name[dState$state %in% states4Bush]
```

```
## [1] "Alabama"      "Arizona"        "Delaware"
## [4] "Georgia"      "Indiana"        "Maine"
## [7] "Missouri"     "Nevada"         "Ohio"
## [10] "Pennsylvania" "Rhode Island"  "Tennessee"
## [13] "Texas"        "West Virginia" "Wyoming"
```

We can then merge the newly created data *dState* with the survey data *dSurvey1* to create a new data *dS1* for our further analysis:

```
# Merge the datasets
dS1 = merge(dSurvey1, dState)
# total observations
dim(dS1)
```

```
## [1] 1611       8
```

```
# How many states
length(unique(dS1$state))
```

```
## [1] 48
```

```
# Print the first 5 observations
head(dS1, 5)
```

```
##   state bush edu age female black state.abbr state.name
## 1     1    0   2   2      1     0         AL    Alabama
## 2     1    1   3   3      1     0         AL    Alabama
## 3     1    1   2   2      0     0         AL    Alabama
## 4     1    1   2   3      0     0         AL    Alabama
## 5     1    1   3   3      1     0         AL    Alabama
```

```
# Print the summary
summary(dS1)
```

```
##      state        bush        edu       age      female    black
## 5      : 190   0   :650   1:191   1:377   0:699   0:1496
## 33     :  86   1   :810   2:573   2:559   1:912   1:  115
## 14     :  82   NA's:151   3:411   3:432
## 44     :  80              4:436   4:243
## 10     :  79
## 36     :  78
## (Other):1016
##    state.abbr          state.name
## Length:1611          Length:1611
## Class :character     Class :character
## Mode  :character     Mode  :character
##
##
##
##
```

9.2 Logistic Regression

Continuing from Chap. 8, let us explore the likelihood of the respondents' preference for Bush with other predictors. Due to the outcome's dichotomous nature, logistic regression should be used to analyze this data. Due to the nested data structure where the respondents are nested within the state, a multi-level logistic regression would be most appropriate to incorporate this nested data structure.

Before we make use of the multi-level logistic regression, let us explore this relationship for a few states to observe whether there are between-state differences from the logistic regression.

As demonstrated in Fig. 9.1, with more education, states with IDs of 1 (i.e., *Alabama*) and 3 (i.e., *Arizona*) seemed to be more in favor of *Bush* and states with IDs of 6 (i.e., *Colorado*) and 7 (i.e., *Connecticut*) were more in favor of *Dukakis*. To investigate this conclusion numerically, we can use the logistic regression to test this conclusion statistically with the following *R* code chunk:

```
# GLM for State #1
glm1 = glm(bush~edu, family="binomial",dS1[dS1$state==1,])
summary(glm1)
```

```
##
## Call:
## glm(formula = bush ~ edu, family = "binomial", data =
##     dS1[dS1$state == 1, ])
##
## Deviance Residuals:
##     Min        1Q    Median        3Q       Max
## -1.6651   -1.3232    0.7585    0.8203    1.0383
##
## Coefficients:
##               Estimate Std. Error z value Pr(>|z|)
## (Intercept)     1.0986     1.1547   0.951    0.341
## edu2           -0.7621     1.2947  -0.589    0.556
## edu3           17.4675  2662.8563   0.007    0.995
## edu4           -0.1823     1.4259  -0.128    0.898
##
## (Dispersion parameter for binomial family taken to be 1)
##
##     Null deviance: 34.162  on 28   degrees of freedom
## Residual deviance: 29.175  on 25   degrees of freedom
##    (2 observations deleted due to missingness)
## AIC: 37.175
##
## Number of Fisher Scoring iterations: 17
```

```
# GLM for State 3
glm3 = glm(bush~edu, family="binomial",dS1[dS1$state==3,])
summary(glm3)
```

```
##
## Call:
## glm(formula = bush ~ edu, family = "binomial", data =
##     dS1[dS1$state == 3, ])
##
## Deviance Residuals:
##      Min        1Q    Median        3Q       Max
## -1.40059  -1.40059   0.00013   0.96954   0.96954
##
## Coefficients:
##               Estimate Std. Error z value Pr(>|z|)
## (Intercept)    -18.57    6522.64  -0.003    0.998
## edu2            19.08    6522.64   0.003    0.998
## edu3            37.13    7292.53   0.005    0.996
## edu4            19.08    6522.64   0.003    0.998
##
```

```
## (Dispersion parameter for binomial family taken to be 1)
##
##       Null deviance: 26.734  on 20   degrees of freedom
## Residual deviance: 21.170  on 17   degrees of freedom
##    (1 observation deleted due to missingness)
## AIC: 29.17
##
## Number of Fisher Scoring iterations: 17
```

```
# GLM for State 6
glm6 = glm(bush~edu, family="binomial",dS1[dS1$state==6,])
summary(glm6)
```

```
##
## Call:
## glm(formula = bush ~ edu, family = "binomial", data =
##     dS1[dS1$state == 6, ])
##
## Deviance Residuals:
##       Min        1Q     Median        3Q        Max
## -1.17741  -0.90052   0.00013   0.88309   1.48230
##
## Coefficients:
##               Estimate Std. Error z value Pr(>|z|)
## (Intercept)  1.857e+01  6.523e+03   0.003    0.998
## edu2         8.309e-09  7.532e+03   0.000    1.000
## edu3        -1.857e+01  6.523e+03  -0.003    0.998
## edu4        -1.926e+01  6.523e+03  -0.003    0.998
##
## (Dispersion parameter for binomial family taken to be 1)
##
##       Null deviance: 19.121  on 13   degrees of freedom
## Residual deviance: 13.183  on 10   degrees of freedom
##    (5 observations deleted due to missingness)
## AIC: 21.183
##
## Number of Fisher Scoring iterations: 17
```

```
# GLM for State 7
glm7 = glm(bush~edu, family="binomial",dS1[dS1$state==7,])
summary(glm7)
```

```
##
## Call:
## glm(formula = bush ~ edu, family = "binomial", data =
##     dS1[dS1$state == 7, ])
##
## Deviance Residuals:
##     Min      1Q   Median      3Q      Max
```

```
## -1.354   -1.177   1.011   1.011   1.177
##
## Coefficients:
##                 Estimate Std. Error z value Pr(>|z|)
## (Intercept)       16.57     2399.54   0.007   0.994
## edu2             -16.16     2399.54  -0.007   0.995
## edu3             -16.57     2399.55  -0.007   0.994
## edu4             -16.57     2399.54  -0.007   0.994
##
## (Dispersion parameter for binomial family taken to be 1)
##
##     Null deviance: 25.864  on 18  degrees of freedom
## Residual deviance: 24.551  on 15  degrees of freedom
##    (4 observations deleted due to missingness)
## AIC: 32.551
##
## Number of Fisher Scoring iterations: 15
```

Based on the intercepts and slopes from these four models, we can observe
that they are all different, which suggests that there is between-state variability. In
addition, the slope parameters in models *glm1* and *glm3* are mostly positive, but
negative in models *glm6* and *glm7*, which suggests that with increased education
level, the probability (measured in log odds) of being in favor of Bush increased for
those respondents in states 1 and 3 but decreased for respondents in states 6 and 7.
We can further notice that these slope parameters are all statistically insignificant
as well as the intercept parameters. This simple illustration leads us to put all the
data together from all 51 states for a multi-level logistic regression to estimate the
between-state and within-state variabilities.

9.3   Fitting Multi-Level Logistic Regression Using *lme4*

The multi-level logistic regression model is one of the several models that can be
categorized as the generalized mixed-effects models (GLMMs). To fit a GLMM, we
use the *R* package *lme4*. The specific function for fitting GLMMs with *lme4* is *glmer*
as described in Appendix B.3. For modeling and optimization of parameter estimation,
the *glmer* function uses an adaptive Gauss–Hermite likelihood approximation
to fit the model to the data.

9.3.1   Random-Intercept GLMM

We now fit the random-intercept GLMM with all the predictors of *edu, age, female,*
and *black* to see which predictors are significantly associated with the outcome,
voting for Bush, as follows:

```
# Load the package
library(lme4)

# Fit a random-intercept GLMM
mS1.intercept = glmer(bush~edu+age+female+
black+(1|state),data=dS1,family =binomial)
summary(mS1.intercept)
```

```
## Generalized linear mixed model fit by maximum likelihood
##     (Laplace Approximation) [glmerMod]
##   Family: binomial  ( logit )
## Formula: bush ~ edu + age + female + black + (1 | state)
##     Data: dS1
##
##      AIC      BIC   logLik deviance df.resid
##   1982.6   2035.4   -981.3   1962.6     1450
##
## Scaled residuals:
##     Min      1Q  Median      3Q     Max
## -1.2940 -1.1506  0.8022  0.8518  2.0360
##
## Random effects:
##  Groups Name        Variance Std. Dev.
##  state  (Intercept) 0.01209  0.11
## Number of obs.: 1460, groups:  state, 47
##
## Fixed effects:
##              Estimate Std. Error z value Pr(>|z|)
## (Intercept)   0.02114    0.20725   0.102   0.9187
## edu2          0.32659    0.18723   1.744   0.0811 .
## edu3          0.37471    0.19837   1.889   0.0589 .
## edu4          0.31957    0.19649   1.626   0.1039
## age2          0.06558    0.14356   0.457   0.6478
## age3          0.04255    0.15209   0.280   0.7797
## age4          0.10361    0.18694   0.554   0.5794
## female1      -0.10659    0.10892  -0.979   0.3278
## black1       -1.35252    0.24322  -5.561 2.68e-08 ***
## ---
## Signif. codes:  0 '***' 0.001 '**' 0.01 '*' 0.05 '.' 0.1 ' ' 1
##
## Correlation of Fixed Effects:
##         (Intr) edu2   edu3   edu4   age2   age3   age4   femal1
## edu2    -0.714
## edu3    -0.706  0.734
## edu4    -0.704  0.742  0.721
## age2    -0.427  0.037  0.021 -0.022
## age3    -0.458  0.063  0.107  0.068  0.565
## age4    -0.468  0.162  0.202  0.199  0.457  0.455
## female1 -0.249 -0.074 -0.032 -0.019  0.003 -0.016 -0.056
## black1  -0.109  0.029  0.017  0.069  0.027  0.033  0.051 -0.047
```

As seen from the model fitting, *female* and *age* are not statistically significant predictors of the probability of *bush* being favored. We then fit a reduced model as follows:

```
# Fit a reduced random-intercept GLMM
mS1.intercept.2 = glmer(bush~edu+black+(1|state),
data=dS1,family =binomial)
summary(mS1.intercept.2)
```

```
## Generalized linear mixed model fit by maximum likelihood
##    (Laplace Approximation) [glmerMod]
##  Family: binomial  ( logit )
## Formula: bush ~ edu + black + (1 | state)
##    Data: dS1
##
##      AIC      BIC   logLik deviance df.resid
##   1975.9   2007.6   -981.9   1963.9     1454
##
## Scaled residuals:
##     Min      1Q  Median      3Q     Max
## -1.2505 -1.1704  0.8200  0.8467  2.0311
##
## Random effects:
##  Groups Name        Variance Std. Dev.
##  state  (Intercept) 0.01048  0.1024
## Number of obs.: 1460, groups:  state, 47
##
## Fixed effects:
##             Estimate Std. Error z value Pr(>|z|)
## (Intercept)  0.02991    0.16242   0.184   0.8539
## edu2         0.30050    0.18400   1.633   0.1024
## edu3         0.35409    0.19303   1.834   0.0666 .
## edu4         0.30382    0.19059   1.594   0.1109
## black1      -1.37106    0.24253  -5.653 1.57e-08 ***
## ---
## Signif. codes:  0 '***' 0.001 '**' 0.01 '*' 0.05 '.' 0.1 ' ' 1
##
## Correlation of Fixed Effects:
##        (Intr) edu2   edu3   edu4
## edu2   -0.860
## edu3   -0.818  0.726
## edu4   -0.835  0.737  0.706
## black1 -0.120  0.019  0.007  0.061
```

In this random-intercept GLMM, the estimated between-state variability in the intercepts is $\hat{\tau}_0 = 0.1024$, which is reported under *Random effects*. The estimated fixed effects for *black* is $\hat{\beta}_4 = -1.371$, which is highly statistically significant with a *p-value = 1.57e−08*. This indicates that African-American respondents were less likely to favor Bush. For education level, respondents with more education (i.e., *2 = high school, 3 = some college,* and *4 = college graduation*) were more likely to favor *Bush*, even though these coefficients were not statistically significant, as demonstrated by $\hat{\beta}_1 = 0.300$, $\hat{\beta}_2 = 0.354$, and $\hat{\beta}_3 = 0.304$ in comparison with the intercept $\hat{\beta}_0 = 0.0299$.

9.3.2 Random-Slope GLMM

Similarly, we can fit the random-slope GLMM with all four predictors as follows:

```
# Fit random-slope GLMM
mS1.slope = glmer(bush~edu+age+female+black+(edu|state),
data=dS1,family=binomial)
summary(mS1.slope)
```

```
## Generalized linear mixed model fit by maximum likelihood
##    (Laplace Approximation) [glmerMod]
##   Family: binomial  ( logit )
## Formula: bush ~ edu + age + female + black + (edu | state)
##      Data: dS1
##
##       AIC       BIC    logLik deviance df.resid
##    1989.1    2089.6    -975.6   1951.1     1441
##
## Scaled residuals:
##     Min       1Q   Median       3Q      Max
## -1.5470  -1.1174   0.7351   0.8379   2.5667
##
## Random effects:
##  Groups Name          Variance Std. Dev. Corr
##  state  (Intercept)   0.5212   0.7220
##         edu2          0.7248   0.8514   -0.99
##         edu3          0.3894   0.6240   -0.78  0.80
##         edu4          0.4162   0.6451   -0.94  0.88  0.84
## Number of obs.: 1460, groups:  state, 47
##
## Fixed effects:
##                Estimate Std. Error z value Pr(>|z|)
## (Intercept)     0.09486    0.25340   0.374    0.708
## edu2            0.24991    0.24947   1.002    0.316
## edu3            0.35408    0.24850   1.425    0.154
## edu4            0.26180    0.24160   1.084    0.279
## age2            0.07750    0.14653   0.529    0.597
## age3            0.05161    0.15522   0.333    0.739
## age4            0.10952    0.19190   0.571    0.568
## female1        -0.12004    0.11177  -1.074    0.283
## black1         -1.45931    0.25131  -5.807 6.37e-09 ***
## ---
## Signif. codes:  0 '***' 0.001 '**' 0.01 '*' 0.05 '.' 0.1 ' ' 1
##
## Correlation of Fixed Effects:
##            (Intr) edu2    edu3    edu4    age2    age3    age4    femal1
## edu2       -0.807
```

```
## edu3      -0.730  0.748
## edu4      -0.780  0.780  0.743
## age2      -0.355  0.022  0.015 -0.017
## age3      -0.384  0.048  0.086  0.064  0.565
## age4      -0.397  0.131  0.175  0.173  0.452  0.450
## female1 -0.204 -0.059 -0.041 -0.022  0.001 -0.020 -0.057
## black1   -0.111  0.049  0.025  0.089  0.029  0.031  0.046  -0.044
## convergence code: 0
```

Since both *age* and *female* are not significant, we can also fit a reduced model as follows:

```
# Fit reduced random-slope GLMM
mS1.slope.2 = glmer(bush~edu+black+(edu|state),
data=dS1,family=binomial)
summary(mS1.slope.2)
```

```
## Generalized linear mixed model fit by maximum likelihood
##    (Laplace Approximation) [glmerMod]
##   Family: binomial  ( logit )
## Formula: bush ~ edu + black + (edu | state)
##    Data: dS1
##
##      AIC      BIC   logLik deviance df.resid
##   1982.7   2062.0   -976.3   1952.7     1445
##
## Scaled residuals:
##     Min      1Q  Median      3Q     Max
## -1.5756 -1.1367  0.7509  0.8403  2.5718
##
## Random effects:
##  Groups Name        Variance Std. Dev. Corr
##  state  (Intercept) 0.5290   0.7274
##         edu2        0.7275   0.8530   -0.99
##         edu3        0.3980   0.6309   -0.80  0.81
##         edu4        0.4363   0.6606   -0.94  0.88  0.85
## Number of obs.: 1460, groups:  state, 47
##
## Fixed effects:
##             Estimate Std. Error z value Pr(>|z|)
## (Intercept)   0.1077     0.2171   0.496    0.620
## edu2          0.2197     0.2470   0.889    0.374
## edu3          0.3268     0.2442   1.338    0.181
## edu4          0.2423     0.2378   1.019    0.308
## black1       -1.4796     0.2506  -5.904 3.55e-09 ***
## ---
## Signif. codes:  0 '***' 0.001 '**' 0.01 '*' 0.05 '.' 0.1 ' ' 1
##
## Correlation of Fixed Effects:
##        (Intr) edu2   edu3   edu4
## edu2  -0.916
## edu3  -0.810  0.747
## edu4  -0.873  0.779  0.739
```

```
## black1 -0.116   0.043   0.017   0.083
## convergence code: 0
```

In this *random-slope* GLMM, the estimated fixed effects for *black* is $\hat{\beta}_4 =$ -1.4796, which is highly statistically significant with *p-value = 3.55e−09*. This again indicates that African-American respondents were less likely to favor Bush. For education level, respondents with more education (i.e., *2 = high school, 3 = some college, and 4 = college graduation*) were more likely to favor *Bush* although this association was not statistically significant, as demonstrated by $\hat{\beta}_1 = 0.2197$, $\hat{\beta}_2 = 0.3268$, and $\hat{\beta}_3 = 0.2423$ in comparison with the intercept $\hat{\beta}_0 = 0.1077$.

This insignificance is largely due to the between-state variability in the intercept and slopes in education level. This can be seen in the following output: $\hat{\tau}_{beta_0} = 0.727$, $\hat{\tau}_{beta_1} = 0.853$, $\hat{\tau}_{beta_2} = 0.631$, and $\hat{\tau}_{beta_3} = 0.661$.

9.3.3   Model Selection

With *glmer*, the nested models (e.g., *mS1.intercept.2* and *mS1.slope.2*) can be compared with one another to determine goodness of fit. Models are compared using the likelihood ratio test with the *anova* function as follows:

```
# Model comparison
anova(mS1.intercept.2, mS1.slope.2)
```

```
## Data: dS1
## Models:
## mS1.intercept.2: bush ~ edu + black + (1 | state)
## mS1.slope.2: bush ~ edu + black + (edu | state)
##                  npar   AIC    BIC  logLik deviance  Chisq Df Pr(>Chisq)
## mS1.intercept.2     6 1975.9 2007.6 -981.94  1963.9
## mS1.slope.2        15 1982.7 2062.0 -976.34  1952.7 11.198  9   0.2624
```

When comparing the models, the null hypothesis is that the two nested models are equivalent. Conversely, a statistically significant result (i.e., $p < 0.05$) indicates that the fit of the two nested models is different. If a statistically significant difference is found, typically, the more complicated model will provide an improvement to the simpler model. In this situation, the random-slope GLMM has a smaller *negative log likelihood* (i.e., 981.94 for random-intercept model vs. 976.34 for random-slope model), but they are not statistically significant based on the likelihood ratio test statistic of $\chi^2 = 11.198$ with degrees of freedom of 9, which produced a *p*-value of 0.2624. Therefore, the random-intercept GLMM is sufficient to fit this data.

In addition, both AIC and BIC results indicate that random-intercept GLMM *mS1.intercept.2* provides a better fit to the data than the random-slope GLMM *mS1.slope.2*, since both AIC and BIC are smaller in the random-intercept GLMM.

9.4 Exercises

1. Follow the analysis in this chapter to analyze the data from other surveys: 9153, 9154, 9155, 9156a, 9156b, 9157, 9158.
2. Analyze the *toenail* data in *R* package *HSAUR2*. (hint: use the following code to get started)

```r
# Load the library
library(HSAUR2)
# Upload the data
data("toenail", package = "HSAUR2")

# Details from the data
summary(toenail)

# Example code for random-intercept model
m1 <- glmer(outcome~treatment*visit+(1|patientID),
                    data=toenail, family=binomial)
```

Chapter 10
Generalized Multi-Level Model for Counts Outcome

In Chap. 8, we reviewed two types of GLMs: logistic regression and Poisson regression. Having discussed generalized linear mixed-effects models for dichotomous outcomes in Chap. 9, we now proceed to the generalized linear mixed-effects model (GLMM) to analyze counts data with multi-level data structures in this chapter.

In this chapter, we will use data reported from Elston et al. (2001) to analyze the observed number of red grouse ticks. This dataset can be found in the *R* package *lme4*. The number of red grouse ticks is *counts data* and will be used to illustrate the multi-level generalized Poisson regression and negative-binomial regression. The multi-level generalized Poisson regression and negative-binomial regression are models included among the generalized linear mixed-effects models (GLMM).

Note to readers: We will use *R* package *lme4* in this chapter. Remember to install this *R* package to your computer using *install.packages("lme4")* and load this package into *R* session using *library(lme4)* before running all the *R* programs in this chapter.

10.1 Data on Red Grouse Ticks

There are two datasets for this analysis. The first dataset is an individual-level data and is named as *grouseticks*. The second dataset contains aggregated data on a brood level and is named as *grouseticks_agg*. The *grouseticks* dataset contains the number of ticks on the heads of 403 red grouse chicks sampled in a field with seven variables as follows:

- *INDEX* is the chick number at the observation level,
- *TICKS* is the number of ticks sampled,
- *BROOD* is their (factor) brood number,
- *HEIGHT* is the height above sea level (meters),

© The Author(s), under exclusive license to Springer Nature Switzerland AG 2021   181
D.-G. (Din) Chen, J. K. Chen, *Statistical Regression Modeling with R*,
Emerging Topics in Statistics and Biostatistics,
https://doi.org/10.1007/978-3-030-67583-7_10

- *YEAR* is the year from 1995 to 1997,
- *LOCATION* is the (factor) geographic location code, and
- *cHEIGHT* is the centered height, derived from HEIGHT.

This dataset can be accessed as follows:

```
# Load the library
library(lme4)

# Load the individual-level data
data(grouseticks)

# name it as dI =  individual-level data
dI = grouseticks

# Check its dimension
dim(dI)
```

```
## [1]  403    7
```

```
# Print the first 5 observations
head(dI,5)
```

```
##    INDEX TICKS BROOD HEIGHT YEAR LOCATION   cHEIGHT
## 1      1     0   501    465   95       32  2.759305
## 2      2     0   501    465   95       32  2.759305
## 3      3     0   502    472   95       36  9.759305
## 4      4     0   503    475   95       37 12.759305
## 5      5     0   503    475   95       37 12.759305
```

Two variables that are included in the aggregated dataset of 118 broods *grouseticks_agg* are:

- *meanTICKS*, which is the mean number of ticks by brood and
- *varTICKS*, which is the variance of number of ticks by brood.

This dataset can be accessed as follows:

```
# Load the aggregated data
data(grouseticks_agg)

# rename it as dA aggregated data
dA = grouseticks_agg

# Check its dimension
dim(dA)
```

```
## [1]  118    7
```

```
# Print the first 5 observations

head(dA,5)
```

```
##   BROOD meanTICKS varTICKS HEIGHT YEAR LOCATION   cHEIGHT
## 1   501 0.0000000 0.000000    465   95       32  2.759305
## 2   502 0.0000000       NA    472   95       36  9.759305
## 3   503 1.2500000 2.250000    475   95       37 12.759305
## 4   504 0.6666667 1.333333    488   95       44 25.759305
## 5   505 0.0000000 0.000000    492   95       47 29.759305
```

Notice that there are some *NAs* in this aggregated dataset. This is because there is only one chick in that particular brood, which results in a *NA* for the variance calculation in *varTICKS*.

10.2 Graphical Analysis

In counts data, the means and variances are typically related, which is why we use Poisson regression since the underlying assumption for Poisson distribution is that the variance is equal to the mean. To illustrate this correlation, we can use the aggregated data to plot the means and variances from all broods as follows:

```
# Plot the mean vs variance
plot(varTICKS~meanTICKS,dA,
     pch=4,xlab="Within-Brood Mean Ticks",
     ylab="Within-Brood Variance")

# add a one-to-one line
abline(0,1, lwd=1, col="red")

# Also add a regression line
m0 = lm(varTICKS~meanTICKS-1,grouseticks_agg)
abline(m0, lwd=3, col="black")
```

Figure 10.1 is similar to Figure 1b in (Elston et al. 2001). In this figure, the thin red line is the one-to-one line indicating that the within-brood variance is equal to the within-brood mean. The additional thick black line is the regression line to fit a slope-only linear regression to identify how many times the within-brood variances are to the within-brood means.

It is clear from this figure that the variances are related to the means from each brood and that the variances are generally larger than the means (i.e., above the one-to-one line). As shown from the simple linear regression, the variances are about 4.325 times as large as the means.

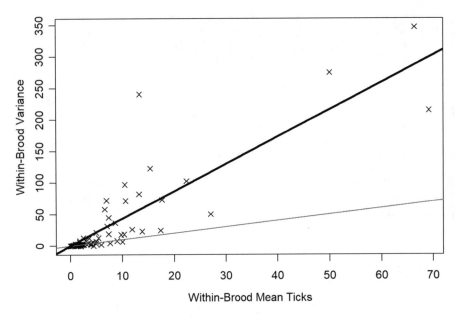

Fig. 10.1 Mean and variance relationship in counts data

To further examine the number of ticks to the height above sea level, we borrow the R code from the R package *lme4* to reproduce Figure 1a from (Elston et al. 2001):

```
## Figure~1a from Elston et al
tvec <- c(0,1,2,5,20,40,80)
pvec <- c(4,1,3)
plot(1+meanTICKS~HEIGHT,dA,pch=pvec[factor(YEAR)],
  log="y",axes=FALSE,xlab="Altitude (m)",
  ylab="Brood Mean Ticks")
axis(side=1); axis(side=2,at=tvec+1,label=tvec)
box()
abline(v=405,lty=2)
```

From Fig. 10.2, we can see that a clear negative relationship exists between the mean number of ticks from each brood to their altitude, which is *YEAR*-specific. Therefore an interaction model between *YEAR* and *HEIGHT* might be more appropriate. We will test this using GLMM.

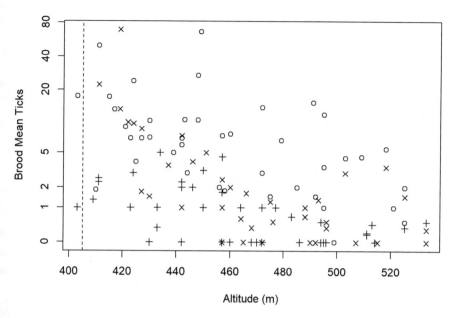

Fig. 10.2 Mean number of ticks to the height above the sea level

10.3 Generalized Linear Mixed-Effects Model with *Poisson* Distribution

We will use *glmer* to fit a series of models to this data and select the most appropriate model with the *R* function *anova*.

10.3.1 Random-Intercept Model

First, we fit the three random-intercept models including both *BROOD* and *LOCA-TION* as follows:

```
set.cont = glmerControl(optimizer = c("Nelder_Mead"))
# Random-intercept model to both brood and location
m0 = glmer(TICKS~YEAR*HEIGHT+(1|BROOD)+(1|LOCATION),
           data=dI,control=set.cont, family="poisson")
summary(m0)

## Generalized linear mixed model fit by maximum likelihood
##    (Laplace Approximation) [glmerMod]
##   Family: poisson  ( log )
## Formula: TICKS ~ YEAR * HEIGHT + (1 | BROOD) + (1 | LOCATION)
##    Data: dI
```

```
## Control: set.cont
##
##      AIC      BIC    logLik deviance df.resid
##   1986.7   2018.7    -985.3   1970.7      395
##
## Scaled residuals:
##     Min      1Q  Median      3Q     Max
## -3.0954 -0.7783 -0.4307  0.6985  5.6365
##
## Random effects:
##  Groups   Name        Variance Std.Dev.
##  BROOD    (Intercept) 0.6350   0.7969
##  LOCATION (Intercept) 0.1873   0.4327
## Number of obs: 403, groups:  BROOD, 118; LOCATION, 63
##
## Fixed effects:
##                Estimate Std. Error z value Pr(>|z|)
## (Intercept)   16.098494   2.524756   6.376 1.81e-10 ***
## YEAR96        -5.881261   3.084985  -1.906   0.0566 .
## YEAR97        -7.118705   3.480662  -2.045   0.0408 *
## HEIGHT        -0.033786   0.005472  -6.175 6.63e-10 ***
## YEAR96:HEIGHT  0.015301   0.006709   2.281   0.0226 *
## YEAR97:HEIGHT  0.013362   0.007592   1.760   0.0784 .
## ---
## Signif. codes:  0 '***' 0.001 '**' 0.01 '*' 0.05 '.' 0.1 ' ' 1
##
## Correlation of Fixed Effects:
##             (Intr) YEAR96 YEAR97 HEIGHT YEAR96:
## YEAR96      -0.757
## YEAR97      -0.664  0.542
## HEIGHT      -0.997  0.756  0.663
## YEAR96:HEIG  0.753 -0.997 -0.539 -0.756
## YEAR97:HEIG  0.659 -0.538 -0.997 -0.662  0.538
## convergence code: 0
```

```
# Random-intercept model to brood only
m1 = glmer(TICKS~YEAR*HEIGHT+(1|BROOD),
           data=dI,control=set.cont,family="poisson")
summary(m1)
```

```
## Generalized linear mixed model fit by maximum likelihood
##   (Laplace Approximation) [glmerMod]
##  Family: poisson  ( log )
## Formula: TICKS ~ YEAR * HEIGHT + (1 | BROOD)
##    Data: dI
## Control: set.cont
##
##      AIC      BIC    logLik deviance df.resid
##   1985.4   2013.4    -985.7   1971.4      396
##
## Scaled residuals:
##     Min      1Q  Median      3Q     Max
## -3.0687 -0.7879 -0.4130  0.6868  5.6328
```

```
##
## Random effects:
##  Groups Name         Variance Std.Dev.
##  BROOD  (Intercept) 0.8167   0.9037
## Number of obs: 403, groups:  BROOD, 118
##
## Fixed effects:
##                  Estimate Std. Error z value Pr(>|z|)
## (Intercept)     16.787578   2.488740   6.745 1.53e-11 ***
## YEAR96          -7.120050   3.184259  -2.236  0.02535 *
## YEAR97          -7.294753   3.585905  -2.034  0.04192 *
## HEIGHT          -0.035218   0.005401  -6.521 7.00e-11 ***
## YEAR96:HEIGHT    0.017937   0.006918   2.593  0.00953 **
## YEAR97:HEIGHT    0.013698   0.007817   1.752  0.07973 .
## ---
## Signif. codes:  0 '***' 0.001 '**' 0.01 '*' 0.05 '.' 0.1 ' ' 1
##
## Correlation of Fixed Effects:
##             (Intr) YEAR96 YEAR97 HEIGHT YEAR96:
## YEAR96      -0.782
## YEAR97      -0.694  0.542
## HEIGHT      -0.997  0.780  0.692
## YEAR96:HEIG  0.779 -0.997 -0.540 -0.781
## YEAR97:HEIG  0.689 -0.539 -0.997 -0.691  0.539
## convergence code: 0
```

```
# Random-intercept model to location only
m2 = glmer(TICKS~YEAR*HEIGHT+(1|LOCATION),
           data=dI,control=set.cont,family="poisson")
summary(m2)
```

```
## Generalized linear mixed model fit by maximum likelihood
##    (Laplace Approximation) [glmerMod]
##  Family: poisson  ( log )
## Formula: TICKS ~ YEAR * HEIGHT + (1 | LOCATION)
##    Data: dI
## Control: set.cont
##
##     AIC     BIC  logLik deviance df.resid
##  2265.5  2293.5 -1125.7   2251.5      396
##
## Scaled residuals:
##    Min     1Q  Median     3Q     Max
## -6.6368 -0.9012 -0.4683  0.9498  5.6721
##
## Random effects:
##  Groups   Name        Variance Std.Dev.
##  LOCATION (Intercept) 0.88     0.9381
## Number of obs: 403, groups:  LOCATION, 63
##
## Fixed effects:
##                  Estimate Std. Error z value Pr(>|z|)
## (Intercept)     14.340863   2.009231   7.137 9.51e-13 ***
```

```
## YEAR96          -1.247534   1.349643  -0.924    0.355
## YEAR97          -8.250198   1.751240  -4.711 2.46e-06 ***
## HEIGHT          -0.029805   0.004412  -6.756 1.42e-11 ***
## YEAR96:HEIGHT  0.005006   0.003108   1.611    0.107
## YEAR97:HEIGHT  0.015627   0.003980   3.927 8.61e-05 ***
## ---
## Signif. codes:  0 '***' 0.001 '**' 0.01 '*' 0.05 '.' 0.1 ' ' 1
##
## Correlation of Fixed Effects:
##             (Intr) YEAR96 YEAR97 HEIGHT YEAR96:
## YEAR96      -0.466
## YEAR97      -0.399  0.581
## HEIGHT      -0.997  0.485  0.414
## YEAR96:HEIG  0.464 -0.998 -0.579 -0.485
## YEAR97:HEIG  0.401 -0.586 -0.998 -0.418  0.585
## convergence code: 0
```

As seen from all three models, we have some convergence issues in the model fitting due to the complexity of the data. At this point, let us keep the default model fitting.

As seen from all the model fittings, the interaction term between *YEAR* and *HEIGHT* is statistically significant. For model selection, we make use of *anova* again to compare these three models as follows:

```
# Compare model 0 to model 1
anova(m0,m1)
```

```
## Data: dI
## Models:
## m1: TICKS ~ YEAR * HEIGHT + (1 | BROOD)
## m0: TICKS ~ YEAR * HEIGHT + (1 | BROOD) + (1 | LOCATION)
##       npar   AIC    BIC  logLik deviance  Chisq Df Pr(>Chisq)
## m1      7 1985.4 2013.4 -985.69   1971.4
## m0      8 1986.7 2018.7 -985.34   1970.7 0.6912  1     0.4058
```

```
# Compare model 0 to model 2
anova(m0,m2)
```

```
## Data: dI
## Models:
## m2: TICKS ~ YEAR * HEIGHT + (1 | LOCATION)
## m0: TICKS ~ YEAR * HEIGHT + (1 | BROOD) + (1 | LOCATION)
##       npar   AIC    BIC   logLik deviance  Chisq Df Pr(>Chisq)
## m2      7 2265.5 2293.5 -1125.75   2251.5
## m0      8 1986.7 2018.7  -985.34   1970.7 280.8  1  < 2.2e-16 ***
## ---
## Signif. codes:  0 '***' 0.001 '**' 0.01 '*' 0.05 '.' 0.1 ' ' 1
```

As seen from the model comparison, *m0* is not statistically different from *m1* since the likelihood ratio statistic is 0.6915 with *p*-value = 0.4056. This indicates that the random-intercept for *LOCATION* may not be needed. In comparing *m0* to

Table 10.1 Estimated fixed effects in random-intercept model

	Estimate	Std. Error	z value	Pr(>|z|)
(Intercept)	16.788	2.489	6.745	0.000
YEAR96	−7.120	3.184	−2.236	0.025
YEAR97	−7.295	3.586	−2.034	0.042
HEIGHT	−0.035	0.005	−6.521	0.000
YEAR96:HEIGHT	0.018	0.007	2.593	0.010
YEAR97:HEIGHT	0.014	0.008	1.752	0.080

m2, we can see that *m0* with the random-intercept for *BROOD* is better than *m2* without *BROOD* to a degree that is statistically significant. Based on this model selection process, we will use the results from *m1*, i.e., the random-intercept for *BROOD*.

With this model, the estimated between-brood variance is $\hat{\tau}^2 = 0.8167$ and the estimated fixed effects can be tabulated in Table 10.1 as follows:

```
knitr::kable(
    round(summary(m1)$coef,3),
    caption = 'Estimated Fixed-Effects in Random-Intercept Model',
    booktabs = TRUE
)
```

We can extract these parameter estimates and add the fitted lines by *YEAR* to Fig. 10.3 as follows:

```
# Get the model estimates
est.par = as.numeric(summary(m1)$coef[,"Estimate"]);est.par
```

```
# Calculate the estimated linear prediction in log-scale
m95 = est.par[1] + est.par[4]*dA$HEIGHT
m96 = (est.par[1]+est.par[2])+ (est.par[4]+est.par[5])*dA$HEIGHT
m97 = (est.par[1]+est.par[3])+ (est.par[4]+est.par[6])*dA$HEIGHT

# # Back-calculate to the estimate the number of ticks
dA$m95 = exp(m95);dA$m96 = exp(m96);dA$m97 = exp(m97)

# sort dA by heights
dA = dA[order(dA$HEIGHT),]
# Re-use the code
par(las=1,bty="l")
tvec <- c(0,1,2,5,20,40,80); pvec <- c(4,1,3)
plot(1+meanTICKS~HEIGHT,dA,pch=pvec[factor(YEAR)],
    log="y",axes=FALSE,xlab="Altitude (m)",ylab="Brood Mean Ticks")
axis(side=1)
axis(side=2,at=tvec+1,label=tvec)
box()
abline(v=405,lty=2)

# add the fitted lines by year
```

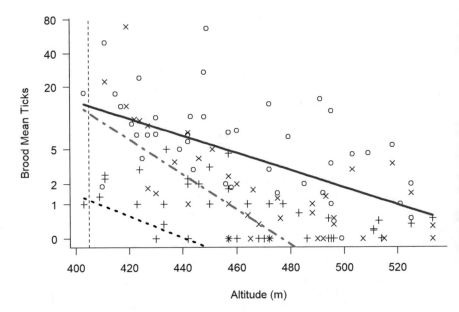

Fig. 10.3 Mean number of ticks to the altitude with the random-intercept model

```
lines((m95)~HEIGHT, dA, lty=4, lwd=3, col="red")
lines((m96)~HEIGHT, dA, lty=1, lwd=3, col="blue")
lines((m97)~HEIGHT, dA, lty=3, lwd=3, col="black")
```

In Fig. 10.3, the red line is for 1995, the blue line is for 1996, and the black line is for 1997. Due to the significant interaction between *YEAR* and *HEIGHT*, the three lines are not parallel with the lowest mean number of ticks in 1997 and the highest in 1995.

10.3.2   Random-Slope Model

We can fit a random-slope model extending the random-intercept model on *BROOD* as follows:

```
# Random-slope model to brood only
m3 = glmer(TICKS~YEAR*HEIGHT+(HEIGHT|BROOD),data=dI,
family="poisson")
summary(m3)
```

```
## Generalized linear mixed model fit by maximum likelihood
##    (Laplace Approximation) [glmerMod]
##   Family: poisson  ( log )
## Formula: TICKS ~ YEAR * HEIGHT + (HEIGHT | BROOD)
```

```
##      Data: dI
##
##      AIC       BIC    logLik deviance df.resid
##    1988.2    2024.2    -985.1   1970.2       394
##
## Scaled residuals:
##     Min       1Q  Median       3Q      Max
## -3.0655  -0.7849  -0.3939   0.6794   5.6326
##
## Random effects:
##  Groups Name          Variance   Std.Dev.  Corr
##  BROOD  (Intercept)   1.529e+01  3.910287
##         HEIGHT        7.666e-05  0.008756  -0.98
## Number of obs: 403, groups:  BROOD, 118
##
## Fixed effects:
##                 Estimate Std. Error z value Pr(>|z|)
## (Intercept)    17.254937   2.563512   6.731 1.69e-11 ***
## YEAR96         -7.200121   3.303315  -2.180   0.0293 *
## YEAR97         -6.930611   3.713310  -1.866   0.0620 .
## HEIGHT         -0.036313   0.005608  -6.475 9.48e-11 ***
## YEAR96:HEIGHT   0.018178   0.007241   2.511   0.0121 *
## YEAR97:HEIGHT   0.012979   0.008156   1.591   0.1115
## ---
## Signif. codes:  0 '***' 0.001 '**' 0.01 '*' 0.05 '.' 0.1 ' ' 1
##
## Correlation of Fixed Effects:
##              (Intr) YEAR96 YEAR97 HEIGHT YEAR96:
## YEAR96       -0.776
## YEAR97       -0.690  0.536
## HEIGHT       -0.998  0.774  0.689
## YEAR96:HEIG   0.773 -0.998 -0.533 -0.775
## YEAR97:HEIG   0.686 -0.532 -0.998 -0.688  0.533
## convergence code: 0
```

Similarly, let us call *anova* to compare this random-slope model to the random-intercept model as follows:

```
# Compare model 1 to model 2
anova(m1,m3)
```

```
## Data: dI
## Models:
## m1: TICKS ~ YEAR * HEIGHT + (1 | BROOD)
## m3: TICKS ~ YEAR * HEIGHT + (HEIGHT | BROOD)
##    npar    AIC    BIC  logLik deviance  Chisq Df Pr(>Chisq)
## m1    7 1985.4 2013.4 -985.69   1971.4
## m3    9 1988.2 2024.2 -985.12   1970.2 1.1496  2     0.5628
```

As seen from the model comparison, these two models are not statistically significantly different since the likelihood ratio statistic is 1.1496 with *p*-value = 0.5628. This indicates that the random-intercept model is sufficient.

10.4 Generalized Linear Mixed-Effects Model with *Negative-Binomial* Distribution

As discussed and illustrated in Fig. 10.1, the variances and means from all *BROOD*s are not the same, which is the basic assumption for Poisson distribution. Therefore, the model fitting with *Poisson* distribution may need to be re-examined.

As discussed in Chap. 8, there are two options for this re-evaluation: the first is to use *quasi-Poisson* and the second is to use *negative binomial*. In the current *lme4*, the *quasi-Poisson* is not available, and we will use the negative-binomial distribution to re-analyze the data.

We will fit a series of models with random-intercept and random-slope to select the most appropriate model to fit this data. The implementation is as follows:

```
# Random-intercept model
m1.nb = glmer.nb(TICKS~YEAR*HEIGHT+(1|BROOD),data=dI)
# Print the summary
summary(m1.nb)
```

```
## Generalized linear mixed model fit by maximum likelihood
##    (Laplace Approximation) [glmerMod]
##  Family: Negative Binomial(3.2943)  ( log )
## Formula: TICKS ~ YEAR * HEIGHT + (1 | BROOD)
##    Data: dI
##
##      AIC       BIC    logLik deviance df.resid
##   1788.9    1820.9    -886.5   1772.9      395
##
## Scaled residuals:
##     Min      1Q  Median      3Q     Max
## -1.4906 -0.6076 -0.2918  0.4732  3.3862
##
## Random effects:
##  Groups Name         Variance Std.Dev.
##  BROOD  (Intercept)  0.7221   0.8497
## Number of obs: 403, groups:  BROOD, 118
##
## Fixed effects:
##                 Estimate Std. Error z value Pr(>|z|)
## (Intercept)    16.939517   2.461839   6.881 5.95e-12 ***
## YEAR96         -7.314650   3.173977  -2.305  0.02119 *
## YEAR97         -7.500960   3.544731  -2.116  0.03434 *
## HEIGHT         -0.035504   0.005342  -6.646 3.01e-11 ***
## YEAR96:HEIGHT   0.018352   0.006895   2.662  0.00778 **
## YEAR97:HEIGHT   0.014167   0.007723   1.834  0.06659 .
## ---
## Signif. codes:  0 '***' 0.001 '**' 0.01 '*' 0.05 '.' 0.1 ' ' 1
##
## Correlation of Fixed Effects:
##            (Intr) YEAR96 YEAR97 HEIGHT YEAR96:
## YEAR96     -0.775
## YEAR97     -0.693  0.538
```

```
## HEIGHT        -0.997  0.773  0.691
## YEAR96:HEIG   0.772 -0.997 -0.536 -0.774
## YEAR97:HEIG   0.689 -0.535 -0.997 -0.690  0.536
## convergence code: 0
```

```
# Random-slope model
m3.nb = glmer.nb(TICKS~YEAR*HEIGHT+(HEIGHT|BROOD),data=dI)
# Print the summary
summary(m3.nb)
```

```
## Generalized linear mixed model fit by maximum likelihood
##   (Laplace Approximation) [glmerMod]
##   Family: Negative Binomial(3.3031)  ( log )
## Formula: TICKS ~ YEAR * HEIGHT + (HEIGHT | BROOD)
##    Data: dI
##
##      AIC      BIC   logLik deviance df.resid
##   1791.7   1831.7   -885.9   1771.7      393
##
## Scaled residuals:
##     Min      1Q  Median      3Q     Max
## -1.4886 -0.6233 -0.2844  0.4679  3.3844
##
## Random effects:
##  Groups Name        Variance  Std.Dev. Corr
##  BROOD  (Intercept) 2.249e+01 4.74225
##         HEIGHT      1.114e-04 0.01055  -0.99
## Number of obs: 403, groups:  BROOD, 118
##
## Fixed effects:
##                Estimate Std. Error z value Pr(>|z|)
## (Intercept)   18.436666   2.931430   6.289 3.19e-10 ***
## YEAR96        -8.745634   3.607409  -2.424  0.01534 *
## YEAR97        -8.297961   3.915229  -2.119  0.03406 *
## HEIGHT        -0.038882   0.006445  -6.033 1.61e-09 ***
## YEAR96:HEIGHT  0.021592   0.007924   2.725  0.00643 **
## YEAR97:HEIGHT  0.016047   0.008585   1.869  0.06161 .
## ---
## Signif. codes:  0 '***' 0.001 '**' 0.01 '*' 0.05 '.' 0.1 ' ' 1
##
## Correlation of Fixed Effects:
##             (Intr) YEAR96 YEAR97 HEIGHT YEAR96:
## YEAR96      -0.806
## YEAR97      -0.705  0.571
## HEIGHT      -0.998  0.804  0.701
## YEAR96:HEIG  0.804 -0.998 -0.567 -0.806
## YEAR97:HEIG  0.705 -0.571 -0.998 -0.704  0.570
## convergence code: 0
```

```
# Model selection
anova(m1.nb, m3.nb)
```

```
## Data: dI
## Models:
```

Table 10.2 Comparison between Poisson and negative binomial

	npar	AIC	BIC	logLik	deviance	Chisq	Df	Pr(>Chisq)
m1	7	1985.380	2013.373	−985.6901	1971.380	NA	NA	NA
m1.nb	8	1788.923	1820.914	−886.4613	1772.923	198.4577	1	0

```
## m1.nb: TICKS ~ YEAR * HEIGHT + (1 | BROOD)
## m3.nb: TICKS ~ YEAR * HEIGHT + (HEIGHT | BROOD)
##         npar    AIC    BIC  logLik deviance  Chisq Df Pr(>Chisq)
## m1.nb      8 1788.9 1820.9 -886.46   1772.9
## m3.nb     10 1791.8 1831.7 -885.87   1771.8 1.1771  2     0.5551
```

From the likelihood ratio test, we see that the random-intercept GLMM is the better choice (p-value $= 0.5551$) than the random-slope GLMM. We can also test whether or not the negative-binomial distribution has a better fit than the simple Poisson distribution as described in Table 10.2 using the R code chunk as follows:

```
knitr::kable(
  anova(m1.nb, m1), caption = 'Comparison between Poisson and Negative-Binomial',
  booktabs = TRUE
)
```

As seen from Table 10.2, the negative-binomial GLMM fit the data much better than the Poisson distribution with a much smaller AIC and BIC. The likelihood ratio statistic is $\chi^2 = 198.458$ with p-value < 0.0001.

10.5   Summary of Model Fitting

In this chapter, we fit a series of models to this data and concluded that the negative-binomial GLMM fit the data much better than the Poisson GLMM based on the likelihood ratio test. This conclusion is consistent with what we observed in Fig. 10.1 where the variances are about 5 times larger than the means.

With this negative-binomial GLMM, the estimated between-brood variance is $\hat{\tau}^2 = 0.7221$ and the estimated fixed effects can be produced in Table 10.3 using the R code chunk as follows:

```
knitr::kable(
  round(summary(m1.nb)$coef,3),
  caption = 'Estimated Fixed-Effects in Negative-Binomial GLMM',
  booktabs = TRUE
)
```

We can also use this negative-binomial GLMM to update Fig. 10.3 from Poisson GLMM as follows:

```
# Get the model estimates
est.par = as.numeric(summary(m1.nb)$coef[,"Estimate"]);est.par
```

Table 10.3 Estimated fixed effects in negative-binomial GLMM

	Estimate	Std. Error	z value	Pr(>|z|)
(Intercept)	16.940	2.462	6.881	0.000
YEAR96	−7.315	3.174	−2.305	0.021
YEAR97	−7.501	3.545	−2.116	0.034
HEIGHT	−0.036	0.005	−6.646	0.000
YEAR96:HEIGHT	0.018	0.007	2.662	0.008
YEAR97:HEIGHT	0.014	0.008	1.834	0.067

```
## [1] 16.93951705 -7.31465007 -7.50095979 -0.03550434  0.01835190
## [6]  0.01416714
```

```
# Calculate the estimated linear prediction in log-scale
m95 = est.par[1] + est.par[4]*dA$HEIGHT
m96 = (est.par[1]+est.par[2])+ (est.par[4]+est.par[5])*dA$HEIGHT
m97 = (est.par[1]+est.par[3])+ (est.par[4]+est.par[6])*dA$HEIGHT

# # Back-calulate to the estimate the number of ticks
dA$m95 = exp(m95);dA$m96 = exp(m96);dA$m97 = exp(m97)
# sort dA by heights
dA = dA[order(dA$HEIGHT),]

# Re-use the code
par(las=1,bty="l")
tvec <- c(0,1,2,5,20,40,80); pvec <- c(4,1,3)
plot(1+meanTICKS~HEIGHT,dA,pch=pvec[factor(YEAR)],
   log="y",axes=FALSE,xlab="Altitude (m)",ylab="Brood Mean Ticks")
axis(side=1)
axis(side=2,at=tvec+1,label=tvec)
box()
abline(v=405,lty=2)

# add the fitted lines by year
lines((m95)~HEIGHT, dA, lty=4, lwd=3, col="red")
lines((m96)~HEIGHT, dA, lty=1, lwd=3, col="blue")
lines((m97)~HEIGHT, dA, lty=3, lwd=3, col="black")
```

Similar to Fig. 10.2, Fig. 10.4 illustrates the significant interaction between *YEAR* and *HEIGHT* with the red line for 1995, blue line for 1996, and black line for 1997. The three lines are not parallel with the lowest mean number of ticks in 1997 and the highest in 1995.

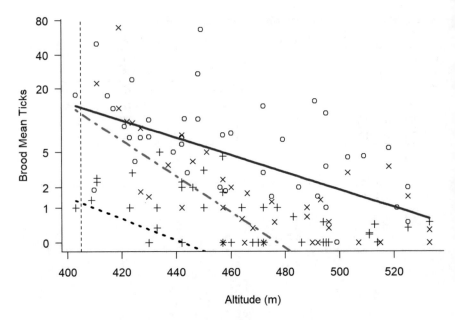

Fig. 10.4 Mean number of ticks to the altitude with negative-binomial GLMM

Appendix A
Motivational Datasets

In this appendix, we list all the datasets used in the book as examples and exercises. We compiled these datasets from different sources so that this book can be self-contained for readers.

A.1   WHO Statistics on Life Expectancy

Compiled from the World Health Organization (WHO) and the United Nations, this dataset is to be used to analyze the factors that affect life expectancy. The data contains 2938 rows and 22 columns from 193 countries. The columns include *country, year, developing status, adult mortality, life expectancy, infant deaths, alcohol consumption per capital, country's expenditure on health, immunization coverage, BMI, deaths under 5 years old, deaths due to HIV/AIDS, GDP, population, body condition, income information,* and *education,* which are defined as follows:

- *Country*: country names;
- *Year*: year of data collected (from 2000 to 2015);
- *Status*: the status of the country on whether it is *Developed* or *Developing*;
- *LifeExpectancy*: life expectancy in age;
- *AdultMortality*: adult mortality rates of both sexes (probability of dying between 15 and 60 years per 1000 population);
- *InfantDeaths*: the number of infant deaths per 1000 population;
- *Alcohol*: alcohol consumption, recorded per capital (15+) consumption (in liters of pure alcohol);
- *PctExpenditure*: expenditure on health as a percentage of Gross Domestic Product per capital;
- *HepatitisB*: hepatitis B (HepB) immunization coverage among 1-year-olds (%);
- *Measles*: the number of reported measles cases per 1000 population;

- *BMI*: average Body Mass Index of entire population;
- *UnderFiveDeaths*: the number of under-five deaths per 1000 population;
- *Polio*: polio (Pol3) immunization coverage among 1-year-olds (%);
- *TotalExpenditure*: general government expenditure on health as a percentage of total government expenditure (%);
- *Diphtheria*: diphtheria tetanus toxoid and pertussis (DTP3) immunization coverage among 1-year-olds (%);
- *HIVAIDS*: deaths per 1000 live births HIV/AIDS (0–4 years);
- *GDP*: Gross Domestic Product per capital (in USD);
- *Population*: population of the country;
- *thin1to19*: the prevalence of thinness among children and adolescents of ages 10–19 (%);
- *Thin5to9*: the prevalence of thinness among children of ages 5–9 (%);
- *Income*: Human Development Index in terms of income composition of resources (index ranging from 0 to 1); and
- *Schooling*: the number of years of schooling (years).

The data is downloaded from https://lionbridge.ai/datasets/10-open-datasets-for-linear-regression/ by Lucas Scott who made available on *10 Open Datasets for Linear Regression*. This dataset is the 10th dataset among these 10 datasets and saved with the name of *WHOLifeExpectancy.csv* for this book.

The dataset can be loaded into *R* as follows:

```
# Read in the WHO life expectancy data
dWHO = read.csv("WHOLifeExpectancy.csv", header=T)
# Check with the dimension of the data: 2938 rows and 22 columns
dim(dWHO)
```

```
## [1] 2938    22
```

```
# Check how many countries in this data
length(unique(dWHO$Country))
```

```
## [1] 193
```

```
# Data Summary
summary(dWHO)
```

```
##    Country                Year            Status
##  Length:2938         Min.   :2000    Length:2938
##  Class :character    1st Qu.:2004    Class :character
##  Mode  :character    Median :2008    Mode  :character
##                      Mean   :2008
##                      3rd Qu.:2012
##                      Max.   :2015
##
##  LifeExpectancy   AdultMortality    InfantDeaths
##  Min.   :36.30    Min.   :   1.0    Min.   :   0.0
```

```
## 1st Qu.:63.10    1st Qu.: 74.0    1st Qu.:   0.0
## Median :72.10    Median :144.0    Median :   3.0
## Mean   :69.22    Mean   :164.8    Mean   :  30.3
## 3rd Qu.:75.70    3rd Qu.:228.0    3rd Qu.:  22.0
## Max.   :89.00    Max.   :723.0    Max.   :1800.0
## NA's   :10       NA's   :10
##    Alcohol         PctExpenditure      HepatitisB
## Min.   : 0.0100   Min.   :    0.000   Min.   : 1.00
## 1st Qu.: 0.8775   1st Qu.:    4.685   1st Qu.:77.00
## Median : 3.7550   Median :   64.913   Median :92.00
## Mean   : 4.6029   Mean   :  738.251   Mean   :80.94
## 3rd Qu.: 7.7025   3rd Qu.:  441.534   3rd Qu.:97.00
## Max.   :17.8700   Max.   :19479.912   Max.   :99.00
## NA's   :194                           NA's   :553
##    Measles            BMI          UnderFiveDeaths
## Min.   :     0.0   Min.   : 1.00   Min.   :   0.00
## 1st Qu.:     0.0   1st Qu.:19.30   1st Qu.:   0.00
## Median :    17.0   Median :43.50   Median :   4.00
## Mean   :  2419.6   Mean   :38.32   Mean   :  42.04
## 3rd Qu.:   360.2   3rd Qu.:56.20   3rd Qu.:  28.00
## Max.   :212183.0   Max.   :87.30   Max.   :2500.00
##                    NA's   :34
##     Polio         TotalExpenditure  Diphtheria
## Min.   : 3.00     Min.   : 0.370    Min.   : 2.00
## 1st Qu.:78.00     1st Qu.: 4.260    1st Qu.:78.00
## Median :93.00     Median : 5.755    Median :93.00
## Mean   :82.55     Mean   : 5.938    Mean   :82.32
## 3rd Qu.:97.00     3rd Qu.: 7.492    3rd Qu.:97.00
## Max.   :99.00     Max.   :17.600    Max.   :99.00
## NA's   :19        NA's   :226       NA's   :19
##    HIVAIDS            GDP             Population
## Min.   : 0.100    Min.   :     1.68  Min.   :3.400e+01
## 1st Qu.: 0.100    1st Qu.:   463.94  1st Qu.:1.958e+05
## Median : 0.100    Median :  1766.95  Median :1.387e+06
## Mean   : 1.742    Mean   :  7483.16  Mean   :1.275e+07
## 3rd Qu.: 0.800    3rd Qu.:  5910.81  3rd Qu.:7.420e+06
## Max.   :50.600    Max.   :119172.74  Max.   :1.294e+09
##                   NA's   :448        NA's   :652
##    Thin1to19       Thin5to9          Income          Schooling
## Min.   : 0.10    Min.   : 0.10    Min.   :0.0000   Min.   : 0.00
## 1st Qu.: 1.60    1st Qu.: 1.50    1st Qu.:0.4930   1st Qu.:10.10
## Median : 3.30    Median : 3.30    Median :0.6770   Median :12.30
## Mean   : 4.84    Mean   : 4.87    Mean   :0.6276   Mean   :11.99
## 3rd Qu.: 7.20    3rd Qu.: 7.20    3rd Qu.:0.7790   3rd Qu.:14.30
## Max.   :27.70    Max.   :28.60    Max.   :0.9480   Max.   :20.70
## NA's   :34       NA's   :34       NA's   :167      NA's   :163
```

A.2   Public Examination Scores in 73 Schools

This dataset can be downloaded from the *Centre for Multilevel Modelling* webpage
at http://www.bristol.ac.uk/cmm/learning/support/datasets/ (named *Exam scores*).

As described in the zipped file, this dataset comes from "The Associated
Examining Board in Guildford", which includes public examination scores from
649 students in 73 centres (i.e., schools or colleges). It has been used to examine
the relationships between candidates' genders and their examination performances.
The available dataset is released only to part of one examination. An analysis of the
complete dataset can be found in Cresswell (1990) and Cresswell (1991).

The dataset contains variables as follows:

* *Centre* is the centre where a candidate comes from with codes 20920-84772,
* *ID* is the candidate identifier with codes 1–5521,
* *Gender* is their gender with boy=0 and girl=1,
* *Written* is the score on written paper with scores 1–144,
* *Course* is the score of coursework evaluated by their teachers with scores 10–108.

The dataset can be loaded into *R* as follows:

```
# Read in the data#
dSci = read.csv("Sci.csv", header=T)
# Check with the dimension of the data
dim(dSci)
```

```
## [1] 1905     5
```

```
# Check how many students in this data
length(unique(dSci$ID))
```

```
## [1] 649
```

```
# Check how many centre in this data
length(unique(dSci$Centre))
```

```
## [1] 73
```

```
# Data Summary
summary(dSci)
```

```
##      Centre            ID            Gender
##   Min.   :20920   Min.    :   1   Min.   :0.0000
##   1st Qu.:60501   1st Qu.:  64   1st Qu.:0.0000
##   Median :68133   Median :  133   Median :1.0000
##   Mean   :62128   Mean    :1037   Mean   :0.5921
##   3rd Qu.:68411   3rd Qu.:  458   3rd Qu.:1.0000
```

```
##    Max.    :84772    Max.    :5521    Max.     :1.0000
##        Written                 Course
##    Min.    :    1.00    Min.    :  10.00
##    1st Qu.:  62.00    1st Qu.:  68.00
##    Median :  75.00    Median :  82.00
##    Mean    :  74.94    Mean      :  79.03
##    3rd Qu.:  89.00    3rd Qu.:  92.00
##    Max.    :144.00    Max.    :108.00
```

The first 5 observations of this dataset can be tabulated in Table A.1 as follows:

```
knitr::kable(
  head(dSci,5), caption = 'Illustration of the First Five Observations',
  booktabs = TRUE
)
```

A.3   Health Behavior in School-Aged Children (HBSC)

Initiated in 1984, the project of "Health Behavior in School-Aged Children (*HBSC*)" had five countries participated in the first round. Since then, the project has grown to a major cross-national survey with more than 40 countries participated to collect data every fourth year among 11-, 13-, and 15-year-old students. The purpose is to collect information about students' health behaviors and well-being. Data is used for research as well as factual basis for policy-making. More information can be found on its webpage (http://www.hbsc.org/).

The data consisted of public, Catholic, and other private school students in grades 5, 6, 7, 8, 9, and 10 or their equivalent in the 50 states and the District of Columbia. Sampling was conducted over three stages: districts, schools, and classes.

Data from 12,642 youths nested within 314 schools are compiled in **hbsc-class.rds** and can be loaded into *R* as follows:

```
# Read in the data
dHBSC = readRDS("hbscclass.rds")
# Check with the dimension of the data: 12,642 rows with 11 columns
dim(dHBSC)
```

```
## [1] 12642        30
```

Table A.1 Illustration of the first five observations

Centre	ID	Gender	Written	Course
20920	16	0	38	22
20920	25	1	60	77
20920	27	1	63	83
20920	31	1	59	95
20920	42	0	27	48

```
# Data Summary
summary(dHBSC)
```

```
##      SCHL_ID            gym               aftsch
##   Min.   :10202   Min.   :-9.000   Min.   :-9.000
##   1st Qu.:33202   1st Qu.: 0.000   1st Qu.: 1.000
##   Median :51204   Median : 0.000   Median : 2.000
##   Mean   :54690   Mean   :-0.865   Mean   : 1.667
##   3rd Qu.:80206   3rd Qu.: 1.000   3rd Qu.: 4.000
##   Max.   :94502   Max.   : 1.000   Max.   : 4.000
##
##      pizza             schsoda            lunch
##   Min.   :-9.000   Min.   :-9.0000   Min.   :-9.000
##   1st Qu.: 1.000   1st Qu.: 1.0000   1st Qu.:-5.000
##   Median : 5.000   Median : 1.0000   Median : 4.000
##   Mean   : 2.615   Mean   : 0.9752   Mean   : 2.208
##   3rd Qu.: 5.000   3rd Qu.: 3.0000   3rd Qu.: 7.000
##   Max.   : 5.000   Max.   : 3.0000   Max.   :10.000
##   NA's   :214
##      trainnut           trainphy            schphy
##   Min.   :-5.0000   Min.   :-5.00000   Min.   :-9.00
##   1st Qu.: 0.0000   1st Qu.: 0.00000   1st Qu.: 1.00
##   Median : 0.0000   Median : 0.00000   Median : 2.00
##   Mean   :-0.1659   Mean   :-0.01353   Mean   : 0.97
##   3rd Qu.: 1.0000   3rd Qu.: 1.00000   3rd Qu.: 2.00
##   Max.   : 1.0000   Max.   : 1.00000   Max.   : 2.00
##
##      schnut            CASEID            DIST_ID
##   Min.   :-9.0000   Min.   :    1    Min.   : -9.00
##   1st Qu.: 1.0000   1st Qu.: 3161   1st Qu.: 25.00
##   Median : 2.0000   Median : 6322   Median : 70.00
##   Mean   : 0.8546   Mean   : 6322   Mean   : 83.75
##   3rd Qu.: 2.0000   3rd Qu.: 9482   3rd Qu.:127.00
##   Max.   : 2.0000   Max.   :12642   Max.   :232.00
##
##      sex               age               white
##   Min.   :-9.0000   Min.   :-9.00    Min.   :0.0000
##   1st Qu.: 0.0000   1st Qu.:12.00    1st Qu.:0.0000
##   Median : 1.0000   Median :13.00    Median :1.0000
##   Mean   : 0.5115   Mean   :12.92    Mean   :0.5206
##   3rd Qu.: 1.0000   3rd Qu.:14.00    3rd Qu.:1.0000
##   Max.   : 1.0000   Max.   :17.00    Max.   :1.0000
##
##      race              body              physact
##   Min.   :-9.000   Min.   :-9     Min.   :-9.000
```

```
##     1st Qu.:  2.000    1st Qu.:  3      1st Qu.:  3.000
##     Median :  2.000    Median :  3      Median :  5.000
##     Mean   :  2.654    Mean   :  3      Mean   :  4.145
##     3rd Qu.:  6.000    3rd Qu.:  4      3rd Qu.:  7.000
##     Max.   :  7.000    Max.   :  5      Max.   :  7.000
##
##          soda              snack              fast
##     Min.   : -9.000    Min.   : -9.0000   Min.   : -9.000
##     1st Qu.:  2.000    1st Qu.: -7.0000   1st Qu.:  2.000
##     Median :  4.000    Median :  3.0000   Median :  4.000
##     Mean   :  3.422    Mean   :  0.3241   Mean   :  3.722
##     3rd Qu.:  6.000    3rd Qu.:  4.0000   3rd Qu.:  5.000
##     Max.   :  7.000    Max.   :  6.0000   Max.   :  7.000
##
##          BMI               diet               frust
##     Min.   : -9.00     Min.   : -9.000    Min.   : -9.000
##     1st Qu.: 15.72     1st Qu.:  1.000    1st Qu.:  1.000
##     Median : 19.35     Median :  1.000    Median :  2.000
##     Mean   : 15.31     Mean   :  1.732    Mean   :  1.631
##     3rd Qu.: 22.54     3rd Qu.:  3.000    3rd Qu.:  3.000
##     Max.   : 48.70     Max.   :  4.000    Max.   :  5.000
##
##          satis             hate               comf
##     Min.   : -9.000    Min.   : -9.000    Min.   : -9.000
##     1st Qu.:  2.000    1st Qu.:  1.000    1st Qu.:  3.000
##     Median :  4.000    Median :  1.000    Median :  4.000
##     Mean   :  2.765    Mean   :  1.105    Mean   :  2.903
##     3rd Qu.:  4.000    3rd Qu.:  2.000    3rd Qu.:  5.000
##     Max.   :  5.000    Max.   :  5.000    Max.   :  5.000
##
##          anger             like               health
##     Min.   : -9.000    Min.   : -9.000    Min.   : -9.000
##     1st Qu.:  1.000    1st Qu.:  3.000    1st Qu.:  1.000
##     Median :  1.000    Median :  4.000    Median :  2.000
##     Mean   :  0.998    Mean   :  2.718    Mean   :  1.832
##     3rd Qu.:  2.000    3rd Qu.:  5.000    3rd Qu.:  2.000
##     Max.   :  5.000    Max.   :  5.000    Max.   :  4.000
##
```

As seen from the summary, this data contains the following variables:

- *Student ID*: named as *CASEID*,
- *School ID*: named as *SCHL_ID*,
- *District ID*: named as *DIST_ID*,
- *Access to Gym*: named as *gym*,
- *After School Programs*: named as *aftsch*,

- *Teacher Training*: named as *train*,
- *Student's sex*: named as *sex*,
- *Student's race*: named as *race*,
- *Physical Activity*: named as *physact*,
- *Body Mass Index*: named as *BMI*, and
- *Overall Health*: named as *health*.

For further explanations,

- *Access to Gym* corresponds to the question, "Have students access to this in unstructured school time? (Breaks, free hours) [Gymnasium, sport hall]" in the School Administrator Questionnaire. The variable is dichotomized with 1 representing "Yes" and 0 representing "No."
- *After School Programs* corresponds to the question, "Does your school organize physical activities during the school day outside physical education classes? [After school]" in the School Administrator Questionnaire. The variable is ordinal with 1 representing "No," 2 representing "Yes, 1–2 days per week," 3 representing "Yes, 3–4 days per week," and 4 representing "Yes, every day."
- *Teacher Training* corresponds to the question, "During the past three years, did the school facilitate staff development (such as workshops, conferences, courses continuing education, or any other kind of in-service training) on the following topics? [Physical Activity—For the Teachers]" in the School Administrator Questionnaire. The variable is dichotomized with 1 representing "Yes" and 0 representing "No."
- *Physical Activity* corresponds to the question, "Over the past 7 days, on how many days were you physically active for a total of at least 60 min per day?" in the Student Questionnaire. The variable is ordinal and ranges from 0 days to 7 days.
- *Body Mass Index* or BMI is a continuous variable based on the respondent's weight and height. It was computed by the HBSC Principal Investigator using the following formula: [Weight (lbs)/[Height (inches) * Height (inches)]] * 703.
- *Overall Health* corresponds to the question, "Would you say your health is …?" in the Student Questionnaire. The variable is ordinal with 1 representing "Excellent," 2 representing "Good," 3 representing "Fair," and 4 representing "Poor."

Acknowledgements HBSC is an international study carried out in collaboration with WHO/EURO. The International Coordinator of the 2005/2006 survey was Prof. Candace Currie and the Data Bank Manager was Prof. Oddrun Samdal. The 2009/2010 survey was conducted by Ronald J. Iannotti in the United States. For details, see http://www.hbsc.org.

A.4 Performance for Pupils in 50 Inner London Schools

This dataset can be downloaded from the *Centre for Multilevel Modelling* webpage at http://www.bristol.ac.uk/cmm/learning/support/datasets/ (named *Junior School Study*).

The original data come from the Junior School Project (Mortimore et al. 1988) and has been used for illustration of multilevel modeling in Goldstein (1987) and Prosser et al. (1991). There are over 1402 students measured over three school years. However, in this dataset, only 1192 students from 49 schools are included with 3236 records in this dataset. The format is as follows:

- *School* is the school ID coded from 1 to 50,
- *Class* is class ID coded from 1 to 4,
- *Gender* is students' gender with *Boy* = 1 and *Girl* = 0,
- *SocialClass* is an indicator for social class coded as I=1, II=2, III nonmanual=3, III manual=4, IV=5, V=6, Long-term unemployed=7, Not currently employed=8, and Father absent=9,
- *RavensTest1* is the Raven's test in year 1 (an ability measure) coded as scores 4–36,
- *ID* is the student ID from 1 to 1402,
- *English* is the English test score coded from 0 to 98,
- *Math* is the mathematics test score coded from 1 to 40,
- *SchoolYear* is the junior school year coded as One=0, Two=1, and Three=2.

The dataset can be loaded into *R* as follows:

```
# Read in data#
dJSP = read.csv("JSP.csv", header=T)
# Check with the dimension of the data
dim(dJSP)
```

```
## [1] 3236      9
```

```
# Check how many school in this data
length(unique(dJSP$School))
```

```
## [1] 49
```

```
# Check how many students in this data
length(unique(dJSP$ID))
```

```
## [1] 1192
```

```
# Data Summary
summary(dJSP)
```

```
##        School              Class              Gender
##    Min.    : 1.00     Min.    :1.000    Min.    :0.0000
##    1st Qu.:14.00      1st Qu.:1.000     1st Qu.:0.0000
##    Median :29.00      Median :1.000     Median :0.0000
##    Mean    :27.09     Mean    :1.531    Mean    :0.4793
##    3rd Qu.:39.00      3rd Qu.:2.000     3rd Qu.:1.0000
##    Max.    :50.00     Max.    :4.000    Max.    :1.0000
##      SocialClass       RavensTest1             ID
##    Min.    :1.00     Min.    : 4.00     Min.    :    1.0
##    1st Qu.:4.00      1st Qu.:21.00      1st Qu.:  361.8
##    Median :4.00      Median :25.00      Median :  703.0
##    Mean    :4.77     Mean    :25.13     Mean    :  701.9
##    3rd Qu.:6.00      3rd Qu.:29.00      3rd Qu.: 1026.0
##    Max.    :9.00     Max.    :36.00     Max.    : 1402.0
##        English             Math             SchoolYear
##    Min.    : 0.00    Min.    : 1.00     Min.    :0.0000
##    1st Qu.:31.00     1st Qu.:22.00      1st Qu.:0.0000
##    Median :54.00     Median :28.00      Median :1.0000
##    Mean    :52.49    Mean    :26.66     Mean    :0.9379
##    3rd Qu.:75.00     3rd Qu.:33.00      3rd Qu.:2.0000
##    Max.    :98.00    Max.    :40.00     Max.    :2.0000
```

A.5   Longitudinal Breast Cancer Post-Surgery Assessment

This is a dataset for longitudinal post-surgery assessment for 405 breast cancer patients, which is used in the paper of Byrne et al. (2008) and also in the book of Byrne (2012) to present latent growth modeling.

The dataset is to present a general overview of measuring change over three waves of data over an 8-month period in individual perceptions of *mood* and *social adjustment* by women who recently (1 month previous) underwent breast cancer surgery (intra-individual change) and to present the longitudinal modeling that measures change across all subjects (inter-individual change).

We will use this dataset for the investigation of 405 Hong Kong Chinese women who recently underwent breast cancer surgery whether exhibited evidence of rate of change in their *mood* and *social adjustment* at 1, 4, and 8 months post-surgery. These women were recruited from six government-funded hospitals in Hong Kong. Women who had linguistic or intellectual difficulties, a currently active Axis I psychiatric diagnosis, or uncontrolled metastatic brain disease were excluded from the study. The women's medical records from the hospitals were screened to determine their eligibility for the study. The research team approached the eligible women about 2–12 days after their surgery. After informed consent was obtained, immediate face-to-face interview was conducted by a member of the research team. The baseline face-to-face interview focused on demographics and measurements of morbidity. The follow-up telephone interviews conducted at post-surgery first month (Wave 1), fourth month (Wave 2), and eighth month (Wave 3) focused on measurements of morbidity and social adjustment. Detailed information on the sample, data collection methods, and sampling procedures is described in Byrne et al. (2008).

In addition, we will use this dataset to demonstrate the addition of *age* and *type* of surgical treatment as possible predictors that may account for any change in the individual growth trajectories (i.e., intercept and slope) of mood and social adjustment.

The dataset is saved with the name of *hkcancer.csv*, and we can examine the structure of this dataset using the following *R* code chunk:

```
# Read in the data
dHK  = read.csv("hkcancer.csv", header=T)
# Check with the dimension of the data: 405 rows with 10 columns
dim(dHK)
```

```
## [1] 405   10
```

```
# Data Summary
summary(dHK)
```

```
##       ID            Mood1           Mood4           Mood8
## Min.   :  1    Min.   :12.00   Min.   :12.00   Min.   :12
## 1st Qu.:102    1st Qu.:16.00   1st Qu.:16.00   1st Qu.:16
## Median :203    Median :20.00   Median :19.00   Median :18
## Mean   :203    Mean   :21.37   Mean   :20.98   Mean   :20
## 3rd Qu.:304    3rd Qu.:25.00   3rd Qu.:25.00   3rd Qu.:23
## Max.   :405    Max.   :44.00   Max.   :43.00   Max.   :47
##                NA's   :35      NA's   :73      NA's   :64
##      SocAdj1         SocAdj4          SocAdj8
## Min.   : 33.00   Min.   : 51.86   Min.   : 64.82
## 1st Qu.: 96.74   1st Qu.: 95.91   1st Qu.: 96.56
## Median :101.06   Median :100.03   Median :100.22
## Mean   :100.91   Mean   :100.42   Mean   :100.60
## 3rd Qu.:106.22   3rd Qu.:105.39   3rd Qu.:105.35
## Max.   :132.00   Max.   :145.20   Max.   :133.14
## NA's   :33       NA's   :69       NA's   :60
##       Age             Age2            SurgTx
## Min.   :28.00    Min.   :0.0000   Min.   :0.0000
## 1st Qu.:44.00    1st Qu.:0.0000   1st Qu.:1.0000
## Median :50.00    Median :0.0000   Median :1.0000
## Mean   :51.59    Mean   :0.4938   Mean   :0.8049
## 3rd Qu.:58.00    3rd Qu.:1.0000   3rd Qu.:1.0000
## Max.   :79.00    Max.   :1.0000   Max.   :1.0000
##
```

We can show the first 10 observations to see the data structure as in Table A.2:

```
knitr::kable(
  head(dHK, 10), caption = 'Illustration of the First Ten Observations',
```

Table A.2 Illustration of the first ten observations

ID	Mood1	Mood4	Mood8	SocAdj1	SocAdj4	SocAdj8	Age	Age2	SurgTx
1	15	NA	NA	95.90625	NA	NA	70	1	1
2	16	25	22	114.88890	105.11110	90.44444	47	0	1
3	37	26	25	80.66667	95.33333	95.33333	47	0	1
4	19	16	15	112.83870	108.58060	99.00000	52	1	1
5	13	16	14	115.00000	105.00000	101.00000	43	0	1
6	21	28	19	106.45160	114.96770	107.51610	34	0	0
7	20	21	16	116.00000	110.00000	106.00000	38	0	1
8	17	NA	NA	93.50000	NA	NA	54	1	1
9	NA	NA	19	NA	NA	107.25000	51	1	0
10	17	15	18	108.90000	114.40000	99.00000	59	1	1

```
    booktabs = TRUE
)
```

It can be seen from the above summary that this dataset has 405 rows (i.e., 405 women) and 10 columns with details:

- *ID* is the identification of the 405 women from 1 to 405,
- *Mood1*, *Mood4*, and *Mood8* are their perceptions of *mood* measured at months 1, 4, and 8 after breast cancer surgery,
- *SocAdj1*, *SocAdj4*, and *SocAdj8* are their perceptions of *social adjustment* measured at months 1, 4, and 8 after breast cancer surgery,
- *Age* represents their age at surgery,
- *Age2* is the dichotomized *age* into two categories of *young* (i.e., $0 = < 51$ years) and *old* (i.e., $1 = > 50$ years), and
- *SurgTx* is the type of surgical treatment (0 = lumpectomy and 1 = mastectomy).

This dataset is presented in the *wide* format (i.e., *person-level*). Each row represents an individual woman and each column represents a recorded measurement/variable for that woman. There are missing values (denoted by *NA*) in the measurements of *mood* and *social adjustment*. Specifically, there are 35, 73, 64, 33, 69, and 60 missing values in *Mood1*, *Mood4*, *Mood8*, *SocAdj1*, *SocAdj4*, and *SocAdj8*, respectively.

For further explanations,

- *Mood* is represented by anxiety and depression symptoms and was measured using the 12-item Chinese Health Questionnaire (CHQ). Women responded to the items of the CHQ on a 4-point Likert scale with anchor points ranging from 0 (Not at all) to 3 (Much more than usual) for those tapping symptoms and from 1 (Much better than usual) to 4 (Much worse than usual) for those tapping positive states. Items were summed up to form a composite score for the scale. Higher scores indicate poorer mood (i.e., more negative symptoms of anxiety and depression).

- *Social Adjustment* is measured using the 33-item Chinese Social Adjustment Scale. Women were asked to indicate perceived change in closeness and comfort levels for a range of five types of social relationships and situations that include family relationships, body image, friend relationships, social activities, and sexuality. For example, one item from the Family subscale asks, *Compared to before you had treatment for your illness, how close have you felt to your children?* All items on the scale are on a 5-point Likert scale with anchors ranging from 1 (Much less than previously) to 5 (A lot more than previously). Higher scores on the scale indicate better social adjustment.

A.6 National Longitudinal Survey of Freshmen (NLSF)

The *NLSF* dataset is compiled from the *National Longitudinal Survey of Freshmen* (conducted from 1999 to 2003) and was developed to provide comprehensive data to test different theoretical explanations for minority underachievement in higher education. NLSF followed a cohort of first-time freshman at selective colleges and universities through their college careers, and equal numbers of Whites, Blacks, Hispanics, and Asians were sampled at each of the 28 participating schools. More details can be found from their webpage at https://nlsf.princeton.edu/.

We will use this dataset for the investigation of 1051 African-American college students whose self-rated *academic effort* was assessed during the second semester of their freshmen (Wave 2), sophomore (Wave 3), junior (Wave 4), and senior (Wave 5) years. At Wave 1, all college students (3924) were given nine statements to assess perceptions of their own and other racial groups; three items addressed intellect, laziness, and perseverance and were coded as internalized negative racial stereotypes based on the previous research by Massey and Owens (2014).

In addition, we will use this dataset to demonstrate the addition of *sex* and *major* (Science, Technology, Engineering, or Mathematics [STEM] versus Non-STEM) as possible predictors that may account for any change in the individual growth trajectories (i.e., intercept and slope) of self-rated academic effort.

The dataset is saved with the name of *nlsfclass.rds*, and we can examine the structure of this dataset using the following *R* code chunk:

```
# Read in the data
nlsf<-readRDS("nlsfclass.rds")
# Check with the dimension of the data: 3924 rows with 17 columns
dim(nlsf)
```

```
## [1] 3924   17
```

```
# Data Summary
summary(nlsf)
```

```
##      caseid            race              sex
```

```
##   Min.   :     8701   Length:3924        Length:3924
##   1st Qu.:24077177   Class :character   Class :character
##   Median :50015148   Mode  :character   Mode  :character
##   Mean   :49571421
##   3rd Qu.:74386787
##   Max.   :99971447
##
##     blkgiveup           blklazy             blkunint
##   Min.   : 1.000   Min.   : 1.000   Min.   : 1.000
##   1st Qu.: 3.000   1st Qu.: 3.000   1st Qu.: 3.000
##   Median : 4.000   Median : 4.000   Median : 4.000
##   Mean   : 6.096   Mean   : 5.795   Mean   : 6.213
##   3rd Qu.: 4.000   3rd Qu.: 4.000   3rd Qu.: 4.000
##   Max.   :98.000   Max.   :98.000   Max.   :98.000
##
##       aef2             pred2               aef3
##   Min.   : 0.000   Min.   :1.000   Min.   : 0.000
##   1st Qu.: 6.000   1st Qu.:1.000   1st Qu.: 6.000
##   Median : 7.000   Median :1.000   Median : 7.000
##   Mean   : 6.828   Mean   :1.651   Mean   : 6.989
##   3rd Qu.: 8.000   3rd Qu.:2.000   3rd Qu.: 8.000
##   Max.   :10.000   Max.   :7.000   Max.   :98.000
##   NA's   :198      NA's   :196     NA's   :504
##       pred3            aef4               pred4
##   Min.   :1.000   Min.   : 0.000   Min.   :1.000
##   1st Qu.:1.000   1st Qu.: 7.000   1st Qu.:1.000
##   Median :1.000   Median : 8.000   Median :2.000
##   Mean   :1.583   Mean   : 7.478   Mean   :1.798
##   3rd Qu.:2.000   3rd Qu.: 8.000   3rd Qu.:2.000
##   Max.   :7.000   Max.   :10.000   Max.   :5.000
##   NA's   :504     NA's   :772      NA's   :851
##       major            aef5              whit0102
##   Min.   :0.0000   Min.   : 0.00   Min.   : 0.00
##   1st Qu.:0.0000   1st Qu.: 7.00   1st Qu.:61.00
##   Median :0.0000   Median : 8.00   Median :71.00
##   Mean   :0.1822   Mean   : 7.57   Mean   :68.64
##   3rd Qu.:0.0000   3rd Qu.: 9.00   3rd Qu.:78.00
##   Max.   :1.0000   Max.   :10.00   Max.   :92.00
##                    NA's   :1453
##     blac0102           uniid
##   Min.   : 3.000   Min.   : 1.000
##   1st Qu.: 5.000   1st Qu.: 4.000
##   Median : 7.000   Median : 8.000
##   Mean   : 8.064   Mean   : 9.496
##   3rd Qu.: 8.000   3rd Qu.:14.000
##   Max.   :87.000   Max.   :28.000
##
```

We can show the first 10 observations for the first 8 variables in Table A.3 to see the data structure:

```
knitr::kable(
  head(nlsf[,1:8], 10),
  caption = 'Illustration of the First Ten Observations',
  booktabs = TRUE
)
```

The number of African-American college students can be generated from:

```
nrow(nlsf[which(nlsf$race=="B"),])
```

```
## [1] 1051
```

We see that there are 1,051 rows (i.e., 1,051 African-American students). This dataset is presented in the *wide* format (i.e., *person-level*). Each row represents an individual college student and each column represents a recorded measurement/variable for that student. The dataset contains 17 columns with details:

- *caseid* is the identification of respondents,
- *uniid* is the identification of the 28 colleges and is computed based on unique values of tuition available in the *institution* dataset,
- *race* is the respondent's race (W=Caucasian/White, A=Asian, B=Black/African-American, and H=Hispanic or Latino),
- *sex* is the respondent's sex (M=Male and F=Female),
- *whit0102* is the percentage of undergraduate White students at the respondent's college, 2001–2002,
- *blac0102* is the percentage of undergraduate Black students at the respondent's college, 2001-2002,
- *major* is the respondent's major and is dichotomized as 0 ("Non-STEM") and 1 ("STEM"),

Table A.3 Illustration of the first ten observations

caseid	Race	Sex	blkgiveup	blklazy	blkunint	aef2	pred2
8701	W	M	4	4	4	9	1
42099	B	M	4	4	4	4	2
44695	H	M	5	4	4	4	2
62115	H	F	4	2	4	10	1
75858	H	M	3	3	3	5	1
76932	A	F	4	4	4	3	1
80484	B	F	3	3	4	7	2
82464	B	F	4	4	3	NA	NA
91100	A	F	3	2	4	6	3
177217	A	F	4	4	4	6	1

- *aef2*, *aef3*, *aef4*, and *aef5* are the respondent's perceptions of *academic effort* ("How hard would you say you have been trying during this past year of college?") measured at Waves 2 through 5 and treated as continuous variables ranging from 0 (No effort) to 10 (Maximum effort),
- *pred2*, *pred3*, and *pred4* are the respondent's perceptions of *campus prejudice* ("How often, if ever, have students in your college classes ever made you feel uncomfortable or self-conscious because of your race or ethnicity?") measured at Waves 2 through 4 and treated as continuous variables ranging from 1 (Never) to 5 (Very often),
- *blklazy* corresponds to the statement "Next, For Each Group I Want To Know Whether You Think They Tend To Be Lazy Or Hardworking [African Americans]." The continuous variable ranges between 1 representing "Hardworking" and 7 representing "Lazy" (reverse coded from the original dataset),
- *blkunint* corresponds to the statement "Next, For Each Group I Want To Know Whether You Think They Tend To Be Unintelligent Or Tend To Be Intelligent [African Americans]." The continuous variable ranges between 1 representing "Intelligent" and 7 representing "Unintelligent" (reverse coded from the original dataset),
- *blkgiveup* corresponds to the statement "Now, For Each Group I Want To Know Whether You Think They Tend To Give Up Easily Or It You Think They Tend To Stick With A Task Until The End [African Americans]." The continuous variable ranges between 1 representing "Stick With A Task Until The End" and 7 representing "Give Up Easily" (reverse coded from the original dataset),

Additional information:

- The *academic effort* and *campus prejudice* variables are time-variant.
- *major* was computed by categorizing 70 majors asked at Wave 5 into a binary variable of STEM and Non-STEM.
- Using the *mutate* function in the *tidyverse* package, *blklazy*, *blkunint*, and *blkgiveup* were aggregated into one continuous variable of internalized negative racial stereotypes based on research established by Massey and Owens (2014). The composite variable *stereo* (ranging from 3 to 21) reflects an increase in the internalization of negative racial stereotypes for each higher score.

Acknowledgements This data is from the National Longitudinal Survey of Freshmen, a project designed by Douglas S. Massey and Camille Z. Charles and funded by the Mellon Foundation and the Atlantic Philanthropies.

A.7   US Population Data

The historical demographics of the United States population data were accessed on April 13, 2020, from https://en.wikipedia.org/wiki/Demographics_of_the_United_States.

The dataset is saved in *USPopulation.csv*, and we can load the data into *R* and print the summary of the data as follows:

```
# Read in the US population data from 1610 to 2010
dUSA = read.csv("uspopulation.csv", header=T)
# Scale the population in millions
dUSA$sPop = dUSA$Pop/1000000
# Data Summary
summary(dUSA)
```

```
##      Decade           Pop                Grate             sPop
## Min.   :1610    Min.   :       350   Min.   :0.0730   Min.   :  0.00035
## 1st Qu.:1710    1st Qu.:    331711   1st Qu.:0.1910   1st Qu.:  0.33171
## Median :1810    Median :   7239881   Median :0.3290   Median :  7.23988
## Mean   :1810    Mean   :  58596031   Mean   :0.5569   Mean   : 58.59603
## 3rd Qu.:1910    3rd Qu.:  92228496   3rd Qu.:0.3703   3rd Qu.: 92.22850
## Max.   :2010    Max.   : 308745538   Max.   :5.5770   Max.   :308.7455
##                                      NA's   :1        4
```

We can show the first 10 observations in Table A.4 to see the data structure:

```
knitr::kable(
  head(dUSA, 10),
  caption = 'First Ten Observations of the US Population',
  booktabs = TRUE
)
```

As seen in this dataset, there are 4 variables:

- *Decade* is the period of decade of that population,
- *Pop* is the population size in that decade,
- *Grate* is the population growth rate, and
- *sPop* is the scaled population size in millions.

As the third most populous country in the world, the population of the United States was estimated at 329,227,746 as of January 28, 2020. As seen from the

Table A.4 First ten observations of the USA population

Decade	Pop	Grate	sPop
1610	350	NA	0.000350
1620	2302	5.577	0.002302
1630	4646	1.018	0.004646
1640	26634	4.733	0.026634
1650	50368	0.891	0.050368
1660	75058	0.490	0.075058
1670	111935	0.491	0.111935
1680	151507	0.354	0.151507
1690	210372	0.389	0.210372
1700	250888	0.193	0.250888

web link, the population growth rates were at over 30% from the early years of 1600s–1880s and then deceased to the range of 20% until 1910s. From 1920s, the population growth rates were less than 20%, and there were about 10% in the decade of 2000.

This data will be used in Chap. 6 to illustrate the nonlinear model and nonlinear regression.

A.8   State of North Carolina COVID-19 Data

Coronavirus disease 2019 (COVID-19) is an infection caused by a novel pathogen named SARS-Cov-2. Spreading worldwide in less than 5 months, the COVID-19 pandemic is a typical example of a global health issue and life-changing infectious disease. In the months, since the first COVID-19 case was reported in the United States on January 22, 2020, many studies have employed different models to reconstruct the epidemic (i.e., the spread of COVID-19 within the United States) and forecast its future trends, from simple growth models to classic susceptible–infected–recovered models. Yet due to the scarcity of available information about the early period of the COVID-19 epidemic, researchers lack sufficient data to construct complex and classic epidemiological models. In this context, the population-based ecological growth model is the preferable option for predicting the epidemic's future trajectory.

To demonstrate the nonlinear growth model of COVID-19, data from the state of North Carolina from the first case reported at March 2, 2020 to July 20, 2020 is downloaded from https://covid19.ncdhhs.gov/dashboard/about-data.

The dataset is saved in *NCCOVID-2020-07-20.csv*, and we can load the data into *R* and print the summary of the data as follows:

```
# Read in the NC Covid-19 data
dNCCOVID = read.csv("NCCOVID-2020-07-20.csv", header=T)
# Data Summary
summary(dNCCOVID)
```

```
##       Day              Date                 Daily
##   Min.   :  1     Length:141           Min.   :   0.0
##   1st Qu.: 36     Class :character     1st Qu.: 191.0
##   Median : 71     Mode  :character     Median : 490.0
##   Mean   : 71                          Mean   : 716.6
##   3rd Qu.:106                          3rd Qu.:1189.0
##   Max.   :141                          Max.   :2481.0
##
##        CumN            DailySpecimen        pctPositive
##   Min.   :     1     Min.   :   1.0       Length:141
##   1st Qu.:  2870     1st Qu.: 195.8       Class :character
```

```
##   Median  : 15045    Median  : 496.0    Mode   :character
##   Mean    : 26864    Mean    : 742.5
##   3rd Qu.: 45102     3rd Qu.:1130.5
##   Max.    :101046    Max.    :2576.0
##                      NA's    :5
##   Hospitalizations NC.Daily.Tests
##   Min.    :  77.0    Min.    :  205
##   1st Qu.: 461.2     1st Qu.: 3310
##   Median : 588.0     Median : 8933
##   Mean    : 641.9    Mean    :11212
##   3rd Qu.: 870.2     3rd Qu.:18041
##   Max.    :1180.0    Max.    :32661
##   NA's    :25        NA's    :14
```

As seen in this dataset, there are 8 variables:

- *Day* denotes the days from the first day of reported COVID-19 cases to the last days (141) of data download,
- *Date* is the actual calendar date,
- *Daily* is the daily counts of COVID-19,
- *CumN* is the cumulative counts,
- *DailySpecimen* is the new cases by specimen date,
- *pctPositive* is the positive test percentage,
- *Hospitalizations* is the number of hospitalization, and
- *NC Daily Tests* is e NC daily tests done.

We can plot the reported daily counts with the following *R* code chunk:

```
# Plot the US population growth rates
par(mar = c(4, 4, .1, .1))
plot(Daily~Day, xlab="Date", ylab="Daily COVID-19 Counts",
     data = dNCCOVID, type = 'b', pch = 19)
```

It can be seen from Fig. A.1 that the daily infection counts of COVID-19 have been increasing from the first day of March 2, 2020. We can also plot the cumulative COVID-19 size using the following *R* code chunk:

```
# Make the plot on population growth
par(mar = c(4, 4, .1, .1))
plot(CumN~Day, xlab="Day", ylab="Cumulative COVID-19 Counts",
     data = dNCCOVID, type = 'b', pch = 19)
```

It can be seen from Fig. A.2 that the cumulative counts of COVID-19 grew from 1 in March 2, 2020 to over 100,000 by July 20, 2020 in a nonlinear (i.e., exponential) fashion. A linear model may not be appropriate to describe this growth model, so a nonlinear regression model should be used.

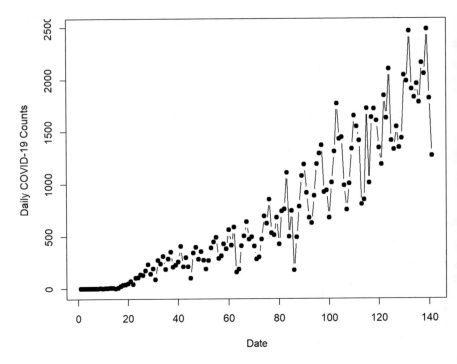

Fig. A.1 North Carolina daily counts

A.9   Political Polls on the 1988 US Presidential Election

The 1988 United States presidential election was the 51st quadrennial presidential election between the Vice President George H. W. Bush as the Republican nominee and the defeated Democratic Massachusetts Governor Michael Dukakis. At the beginning of the campaign, Dukakis was leading all the polls, but Bush ran a very aggressive campaign with focuses on the economy with the continuation of President Ronald Reagan's policies. In the September's opinion poll, Bush led the poll and won the election by a substantial margin in both the popular and the electoral vote. No candidate since 1988 has surpassed Bush's success of the electoral or popular vote.

This dataset was compiled and used as an example in the book *Data Analysis Using Regression and Multilevel/Hierarchical Model* (Gelman and Hill 2007) (http://www.stat.columbia.edu/~gelman/arm/). We downloaded the data from their webpage: http://www.stat.columbia.edu/~gelman/arm/examples/election88.

Details about this dataset can be found from the book. In this dataset, there are 13,544 observations and 10 variables:

- *org* with only value of *cbsnyt* indicating that this survey was done by CBS New York Time,

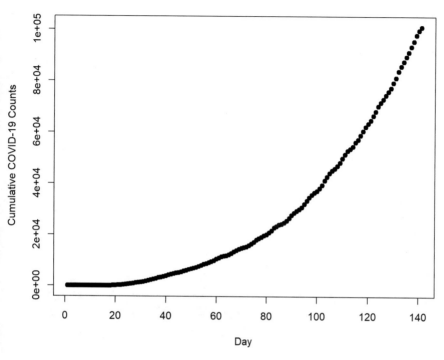

Fig. A.2 North Carolina cumulative counts

- *year* with values from 1 to 7 corresponding to the 8 *survey*s conducted with survey IDs of 9152, 9153, 9154, 9155, 9156a, 9156b, 9157, and 9158.

The number of voters in each survey can be summarized in Table A.5 using the following *R* code chunk:

```
# Read in the polls data
dPoll = read.csv("polls.csv", header=T)
mtab  = aggregate(dPoll$org, list(year=dPoll$year, survey=dPoll$survey), length)
mtab  = data.frame(year=mtab$year,survey=mtab$survey, Num = mtab$x)
knitr::kable(
    mtab, caption = 'Number of Voters in each year and survey',
    booktabs = TRUE
)
```

- *bush* with values 0 or 1 where *bush = 1* indicates that the voters are in favor of *George H. W. Bush* and otherwise in favor of Michael Dukakis if *bush = 0*,
- *state* with values 1 to 51 (including the District of Columbia),
- *edu* is the education level with 4 categories where *1 = no high school, 2 = high school, 3 = some college,* and *4 = college graduation,*
- *age* is their age with 4 categories where *1 = people aged 18–29 years, 2 = people aged 30–44 years, 3 = people aged 45–64 years,* and *4 = people over 65 years old,*

Table A.5 Number of voters
in each year and survey

Year	Survey	Num
1	9152	1611
2	9153	1653
3	9154	1833
4	9155	1943
5	9156a	684
5	9156b	1478
6	9157	2149
7	9158	2193

- *female* is the indicator for gender with *1 = female* and *0 = male*,
- *black* is the indicator for race with *1 = African-American* and *0 = other*, and
- *weight* is the survey weight.

The dataset can be loaded into *R* as follows:

```
# Read in the polls data#
dPoll = read.csv("polls.csv", header=T)
# Check the dimension of the data
dim(dPoll)
```

```
## [1] 13544     10
```

```
# Show the first 6 observations
head(dPoll)
```

```
##        org year survey bush state edu age female black weight
## 1 cbsnyt    1   9152    1     7   2   2      1     0   1403
## 2 cbsnyt    1   9152    1    33   4   3      0     0    778
## 3 cbsnyt    1   9152    0    20   2   1      1     0   1564
## 4 cbsnyt    1   9152    1    31   3   2      1     0   1055
## 5 cbsnyt    1   9152    1    18   3   1      1     0   1213
## 6 cbsnyt    1   9152    1    31   4   2      0     0    910
```

We will use this dataset to illustrate the logistic regression and multi-level logistic regression to model the probability that a respondent would choose the Republican candidate for president.

Appendix B
R Packages

There are two popular *R* packages for longitudinal and multi-level data analysis. One is the *nlme* for *Linear and Nonlinear Mixed-Effects Models* created by Jose Pinheiro, Douglas Bates, and their colleagues. The online help on *nlme* can be very helpful, but their book (Pinheiro and Bates 2000) explained the statistical theory as well as the *R* and *S* implementations. The other *R* package is *lme4* for *Linear Mixed-Effects Models using "Eigen" and S4* created by Douglas Bates and his colleagues. The *nlme* is mostly used for either *linear* or *nonlinear* mixed-effects modeling for continuous outcome. The *lme4* can be used for *linear* and *nonlinear* mixed-effects modeling with both continuous data and non-normal data (including binary, multinomial, and counts data).

In this book, we will use the package *nlme* where the function *lme* for linear mixed-effects modeling and *nlme* for nonlinear mixed-effects modeling for continuous data in Chaps. 2–7. The function *glmer* in *lme4* will be used for generalized linear mixed-effects model for binary and counts data in Chaps. 9–10.

Therefore, you need to load this *R* package before you practice the code in this book as follows:

```
# Load the nlme package
library(nlme)
# Load the lme4 package
library(lme4)
```

For any questions on datasets and functionalities included in the packages, we can call the *help* to check them all in the packages as follows:

```
# Help on the nlme package
library(help=nlme)
# Help on the lme4 package
library(help=lme4)
```

© The Author(s), under exclusive license to Springer Nature Switzerland AG 2021
D.-G. (Din) Chen, J. K. Chen, *Statistical Regression Modeling with R*,
Emerging Topics in Statistics and Biostatistics,
https://doi.org/10.1007/978-3-030-67583-7

To be self-contained in the book, we briefly explain some of the functions used in the book.

B.1   Function *lme* in *R* Package *nlme* for Linear Mixed-Effects Modeling

The function *lme* from the package *nlme* is a generic function to fit a linear mixed-effects model allowing for nested random effects. The within-group errors are allowed to be correlated and/or have unequal variances. We will use this function for chapters on longitudinal data analysis and multi-level modeling with linear mixed-effects models for continuous data in Chaps. 3–6.

The usage for *lme* is as follows:

```
lme(fixed, data, random, correlation, weights, subset, method,
    na.action, control, contrasts = NULL, keep.data = TRUE)
```

where

- *fixed* is a two-sided linear formula object for the fixed-effects part of the model,
- *data* is an optional dataframe containing the variables in fixed, random, correlation, weights, and subset,
- *random* is to specify the random effects of the model,
- *correlation* is an optional *corStruct* object describing the within-group correlation structure with default to NULL indicating within-group correlations,
- *weights* is an optional *varFunc* object describing the within-group heteroscedasticity structure with default to NULL indicating homoscedastic within-group errors,
- *subset* is an optional expression indicating that the subset of the data to be used in the fit with all observations is included by default,
- *method* is a character string with *REML* for fitting by maximizing the restricted log-likelihood and *ML* for the log-likelihood is maximized, and the default is *REML*,
- *na.action* is an indicator for NAs that the default action of *na.fail* causes *lme* to print an error message and terminate if there are any incomplete observations,
- *control* is a list of control values for the estimation algorithm to replace the default values returned by the function *lmeControl*, and
- *contrasts* is an optional list which can be seen from the *contrasts.arg* of *model.matrix.default*.

B.2   Function *nlme* in *R* Package *nlme* for Nonlinear Mixed-Effects Modeling

The function *nlme* from the package *nlme* is a generic function to fit a non-linear mixed-effects model allowing for nested random effects. The within-group errors are allowed to be correlated and/or have unequal variances. We will use this function for Chap. 8 on nonlinear mixed-effects models.

The syntax for *nlme* is as follows:

```
nlme(model, data, fixed, random, groups, start, correlation, weights,
     subset, method, na.action, naPattern, control, verbose)
```

where

- *model* is a nonlinear model formula with the response on the left of a ~ operator and an expression involving parameters and covariates on the right,
- *data* is an optional dataframe containing the variables named in model, fixed, random, correlation, weights, subset, and naPattern,
- *fixed* is a two-sided linear formula to specify the *fixed effects* in the model,
- *random* is to specify the *random effects* in the model,
- *groups* is an optional one-sided formula to specify the *groups* in the model,
- *start* is an optional numeric vector or list to specify the initial estimates for the fixed effects and random effects,
- *correlation* is an optional *corStruct* object describing the within-group correlation structure,
- *weights* is an optional varFunc object or one-sided formula describing the within-group heteroscedasticity structure,
- *subset* is an optional expression indicating the subset of the rows of data that should be used in the fit,
- *method* is a character string for *REML* to fit by maximizing the restricted log-likelihood or for *ML* to fit by maximizing the log-likelihood with defaults to *ML*,
- *na.action* is a function that indicates what should happen when the data contain NAs. The default action (na.fail) causes *nlme* to print an error message and terminate if there are any incomplete observations,
- *naPattern* is an expression or formula object, specifying which returned values are to be regarded as missing,
- *control* is a list of control values for the estimation algorithm to replace the default values returned by the function nlmeControl with defaults to an empty list, and
- *verbose* is an optional logical value. If TRUE information on the evolution of the iterative algorithm is printed. Default is FALSE.

B.3   Function *glmer* in *R* Package *lme4* for Generalized Linear Mixed-Effects Modeling

The function *glmer* in *R* package *lme4* is used to fit a generalized linear mixed-effects model (GLMM) with both fixed effects and random effects.
The general usage is as follows:

```
glmer(formula, data = NULL, family = gaussian, control = glmerControl(),
    start = NULL, verbose = 0L, nAGQ = 1L, subset, weights, na.action,
    offset, contrasts = NULL, mustart, etastart,
    devFunOnly = FALSE, ...)
```

where

- *formula* is a two-sided linear formula object describing both the fixed effects and random effects, with the response on the left of a ~ operator and the terms, separated by + operators, on the right. Random effects terms are distinguished by vertical bars ("l") separating expressions for design matrices from grouping factors.
- *data* is an optional dataframe containing the variables named in formula.
- *family* is a GLM family.
- *control* is a list containing control parameters.
- *start* is a named list of starting values for the parameters in the model.
- *verbose* is integer scalar. If > 0, verbose output is generated during the optimization of the parameter estimates. If > 1, verbose output is generated during the individual penalized iteratively reweighted least squares (PIRLS) steps.
- *nAGQ* is integer scalar—the number of points per axis for evaluating the adaptive Gauss–Hermite approximation to the log-likelihood; defaults to 1, corresponding to the Laplace approximation. The values greater than 1 produce greater accuracy in the evaluation of the log-likelihood at the expense of speed. A value of zero uses a faster but less exact form of parameter estimation for GLMMs by optimizing the random effects and the fixed-effects coefficients in the penalized iteratively reweighted least squares steps.
- *subset* is an optional expression indicating the subset of the rows of data that should be used in the fit.
- *weights* is an optional vector of *prior weights* to be used in the fitting process.
- *na.action* is a function that indicates what should happen when the data contain NAs.)) strips any observations with any missing values in any variables.
- *offset* is used to specify an a priori known component to be included in the linear predictor during fitting.
- *contrasts* is an optional list to specify contrasts.
- *mustart* is optional starting values on the scale of the conditional mean, as in glm.
- *etastart* is optional starting values on the scale of the unbounded predictor as in glm.
- *devFunOnly* is logical value to return only the deviance evaluation function.

Bibliography

Bates, D.M., Watts, D.G.: Nonlinear Regression Analysis and Its Applications. Wiley, London (1998)

Byrne, B.M.: Structural Equation Modeling with Mplus. Routledge, New York (2012)

Byrne, B.M., Lam, W.W.T., Fielding, R.: Measuring pattern of change in personality assessments: an annotated application of latent growth curve modelling. J. Pers. Assess. **90**, 536–546 (2008)

Chambers, J.M., Hastie, T.J.: Statistical Models in S. Wadsworth & Brooks/Cole, New York (1992)

Chen, D.G., Chen, X., Chen, J.K.: Reconstructing and forecasting the covid-19 epidemic in the United States using a 5-parameter logistic growth model. Global Health Res. Policy **5**, 25 (2020). https://doi.org/10.1186/s41256-020-00152-5

Cresswell, M.: Gender Effects in GCSE: Some Initial Analyses. Associated Examining Board Research Report RAC/517 (1990)

Cresswell, M.: A multilevel bivariate model. In: Prosser, R., Rasbash, J., Goldstein, H. (eds.) Data Analysis with ML3. Institute of Education, University of London (1991)

Diggle, P., Heagerty, P., Liang, K.-Y., Zeger, S.: Analysis of Longitudinal Data, 2nd edn. Oxford University Press, London (2002)

Dobson, A.J., Barnett, A.G.: An Introduction to Generalized Linear Models, 3rd edn. Chapman & Hall/CRC Texts in Statistical Science (2018)

Dunn, P.K., Smyth, G.K.: Generalized Linear Models With Examples in R. Springer, New York (2018)

Elston, D.A., Moss, R., Boulinier, T., Arrowsmith, C., Lambin, X.: Analysis of aggregation, a worked example: numbers of ticks on red grouse chicks. Parasitology **122**(5), 563–569 (2001)

Finch, W.H., Bolin, J.E., Kelley, K.: Multilevel Modeling Using R, 2nd edn. CRC Press: Taylor & Francis Group, New York (2019)

Fox, J., Weisberg, S.: Applied Regression Analysis and Generalized Linear Models, 3rd edn. Sage Publications, New York (2016)

Fox, J., Weisberg, S.: An R Companion to Applied Regression, 3rd edn. Sage Publications, New York (2019)

Gelman, A., Hill, J.: Data Analysis Using Regression and Multilevel/Hierarchical Models. Cambridge University Press, Cambridge (2007)

Goldstein, H.: Multilevel Models in Educational and Social Research. Oxford University Press, London, Griffin; New York (1987)

Levy, P.S., Lemeshow, S.: Sampling of Populations: Methods and Applications, 4th edn. Wiley, New York (2008)

Malthus, T.R.: An Essay on the Principle of Population. Oxford World's Classics reprint (1798)

Massey, D.S., Owens, J.: Mediators of stereotype threat among black college students. Ethn. Racial Stud. **37**(3), 557–575 (2014)

McCullagh, P., Nelder, J.A.: Generalized Linear Models, 2nd edn. Chapman and Hall, London (1995)

Mortimore, P., Sammons, P., Stoll, L., Lewis, D., Ecob, R.: School Matters, the Junior Years. Wells, Open Books (1988)

Nash, J.C.: Nonlinear Parameter Optimization Using R Tools. Wiley, West Sussex (2014)

Pinheiro, J.C., Bates, D.M.: Mixed-Effect Models in S and SPLUS. Springer, New York (2000)

Prosser, R., Rasbash, J., Goldstein, H.: ML3 Software for Three-level Analysis, Users' Guide for V2. Institute of Education, University of London (1991)

Snijders, T.A.B., Bosker, R.J.: Multilevel Analysis: An Introduction to Basic and Advanced Multilevel Modeling, 2nd edn. Sage Publishers, London (2012)

Thompson, S.K.: Sampling, 3rd edn. Wiley, New Jersey (2012)

Verhulst, P.: Notice sur la loi que la population suit dans son accroissement. Correspondence Math. Phys. **10**, 113–121 (1838)

Wilson, J.R., Chen, D.G.: Fundamental Statistical Analytics for Health Data: Using R/SAS/STATA/SPSS, vol. 1. Springer, New York (2021)

Wilson, J.R., Lorenz, K.A.: Modeling Binary Correlated Responses Using SAS, SPSS and R. Springer, New York (2015)

Wilson, J.R., Vazquez, E., Chen, D.G.: Marginal Models in Analysis of Correlated Binary Data with Time Dependent Covariates. Springer. New York (2020)

Xie, Y.: *Bookdown: Authoring Books and Technical Documents with R Markdown* (2019). R package version 0.13

Index

A
Akaike Information Criterion (AIC), 13–20, 55–57, 60–68, 80–85, 88, 103–107, 110–112, 137–141, 152, 159, 161, 162, 172–179, 186–188, 191–194

ANOVA, 11, 29, 64, 67–68, 88, 104, 111, 113, 136, 139, 140, 179, 185, 188, 191, 193, 194

B
Between-cluster variance, 30–32, 38, 54, 56, 58, 77, 78

Between-cluster variation, 31, 32, 39

BIC, 14, 55–57, 60–68, 77, 80–88, 103–107, 110–113, 138–141, 175–179, 186–188, 191–194

Binary data, 146, 165

Breast cancer, 91–93, 206–208

C
Chi-square, 63–68, 83

Cluster sampling, 27–29

Confidence intervals, 69–70, 79, 81, 101, 102, 111, 149

Correlation coefficient, 52, 98, 103, 108, 112

Counts data, vi, vii, 143, 155–163, 181, 183, 184, 215, 219

Covariance structures, 109–113

COVID-19, 115, 117, 214–216

D
Datasets
  bioChemists, 155, 156, 158–160, 162
  coronaryArtery.csv, 144
  grouseticks, 181–183
  grouseticks_agg, 181–183
  hbscclass.rds, 201
  hkcancer.csv, 92, 207
  JSP.csv, 72, 205
  loblolly, 131–134, 137–139, 141
  NCCOVID-2020-07-20.csv, 214
  nlsfclass.rds, 209
  polls.csv, 166, 217, 218
  Sci.csv, 34, 46, 200
  toenail, 180
  USPop, 129
  uspopulation.csv, 123, 213
  WHOLifeExpectancy.csv, 1, 198

Degrees of freedom, 5, 6, 10, 11, 14, 15, 20, 44, 54, 59, 100, 101, 126, 134, 146, 152, 157, 159, 161, 162, 172–174, 179

Diagnostics, 5, 7, 8, 21–27, 143–145

Dichotomous data, 143–154, 165–180

Dichotomous outcome, 143–154, 165–180

Dispersion parameter, 152, 159–162, 172–174

F
Fixed-effects, 38, 40, 46, 54, 56, 58, 60, 61, 69, 76, 77, 80, 81, 83, 141, 189, 194, 220, 222

F-test, 6, 11

© The Author(s), under exclusive license to Springer Nature Switzerland AG 2021
D.-G. (Din) Chen, J. K. Chen, *Statistical Regression Modeling with R*,
Emerging Topics in Statistics and Biostatistics,
https://doi.org/10.1007/978-3-030-67583-7

Printed in the United States
by Baker & Taylor Publisher Services